Home Studio Setup

Home Studio Setup

Everything You Need to Know from Equipment to Acoustics

Ben Harris

AMSTERDAM • BOSTON • HEIDELBERG • LONDON • NEW YORK
OXFORD • PARIS • SAN DIEGO • SAN FRANCISCO
SINGAPORE • SYDNEY • TOKYO

Focal Press is an imprint of Elsevier

Focal Press is an imprint of Elsevier
30 Corporate Drive, Suite 400, Burlington, MA 01803, USA
Linacre House, Jordan Hill, Oxford OX2 8DP, UK

Library of Congress Cataloging-in-Publication Data
Application submitted

British Library Cataloguing-in-Publication Data
A catalogue record for this book is available from the British Library.

ISBN: 978-0-240-81134-5

For information on all Focal Press publications
visit our website at www.elsevierdirect.com

09 10 11 12 13 5 4 3 2 1

Printed in the United States of America

This book is dedicated to my wife and children for their patience and support; and to all of the passionate home recordists desiring clear, accurate, and pertinent information about home recording.

Contents

SECTION 2 • EQUIPMENT

SECTION 3 • RECORDING TECHNIQUES

Introduction

Today home recording is an exploding hobby around the world for multiple reasons. Equipment is affordable, gear is cheap, and the price of recording equipment has gone down (wait ... that's really just one reason). The fact that recording equipment has entered an affordable range is the one and only reason that home recording has grown in popularity as a hobby in the last 20 years. Recording used to be only available, in good quality, through high-end expensive recording studios. Thirty years ago rates were usually at least $60 per hour, whether you were recording an album or archiving your 80-year-old father's war stories for posterity's sake. Then in 1991 the ADAT, an eight-channel digital recorder in the range of $3000, was released. The professional-level home studio was born. Nearly 20 years later the ADAT has gone the way of the dodo, high-end studios have lost their monopoly, and professional-level products are coming out of project and home studios.

This new dynamic has come with fear, prejudice, and excitement. As their monopoly has crumbled, many members of the recording industry have despised the rise of home studios. Equipment manufacturers have had a heyday, making new toys and exciting gear so quickly that you can hardly keep up to date. Meanwhile the hobbyist has grown to desire higher and higher recording quality. Thousands of books, courses, and magazine articles bombard the home-recording community with an array of excellent, limited, and sometimes inaccurate information. It is difficult as an uneducated user to sift through it all.

It also doesn't help that serious hobbyists often find that they are not included in the target audience of most books and articles. I've seen that there are two general audiences targeted, ostracizing the rest of you. One target is the 14-year-old kid who isn't serious, doesn't have much of budget, and is all about the hype. The other is the seasoned home recording hobbyist who still has an analog tape machine or ADAT and has no desire to record on a computer.

What about everybody else? What about the doctor who wants to have a cool home studio to record his band on weekends? What about the retired stockbroker who desires higher-quality recording but doesn't want a second career in audio? What about the biologist who wants to keep progressing as a songwriter by producing albums from her basement? What about the hobbyist who wants to know everything that the pros know but doesn't want to become a pro? No book out there gives the big picture and lays it out straight. Nobody describes all the levels of quality and professionalism and says, "Here you go, pick what level you want to be." This book does all that and more.

In three straightforward sections this book demonstrates how to build and acoustically treat a home-recording space. It explains the ins and outs of recording equipment so that the buyer is educated. It covers tried-and-true recording techniques and concepts.

This book is also accompanied by a companion Website found at theDAWstudio.com. This site includes additional acoustic tips, current equipment reviews, and more recording techniques that all add to the knowledge presented here in the book.

As an audio engineer and educator, I feel that the recording industry should embrace the new home studio dynamic. There will always be a place for professionals, and the more thoroughly the hobbyist is educated, the more clear that place becomes.

SECTION 1
Acoustics

CHAPTER 1
Introduction to Acoustics

Acoustics can seem like such a basic thing that you don't need to worry about it, or it can be so confusing that the thought of dealing with it keeps you up at night. The next few chapters cover acoustics issues related to the home recording studio on a broad introductory level. The goal is to give you a solid, basic understanding of how sound functions, what common problems occur, and how to remedy those problems. If you want to learn more detailed information about acoustics, there are many wonderful complete books on the topic. These chapters take you to the level of knowledge of most professional recording engineers (users, not designers of studios). All this information will prove useful in understanding microphone placement, more critically listening to speakers, and acoustically treating any space.

Acoustics are very important in the recording studio for two main reasons. First, to record an acoustic instrument properly with a microphone, the mic needs to react well, acoustically, in its space. Second, there must be proper acoustics during the speaker playback so that the listener can hear what is really happening in the mix, to confidently make changes to the audio. These two factors can easily be compared to factors in other artistic fields. For example, improper lighting at a photo shoot could prove disastrous to the final product, no matter how much processing is done. Colors for a bright and exuberant candy commercial might come out looking dull and boring if processed on a black-and-white video monitor. These are just a few of the reasons that acoustics are so important in any recording studio. The key to effective acoustic treatment is first to understand how sound waves interact in an environment—in this case, your room.

FUNDAMENTALS OF ACOUSTICS

How Sound Travels Through Air

They are called *sound waves* for a reason. They act very similar to waves in a pool of water. Instead of moving water back and forth, sound waves move air. Think about it this way: If you are at one end of a pool with your hand in the water and your friend drops a rock in the other end of the pool, when the wave eventually gets to your hand, is the water touching your hand the same water molecules that the rock touched when it entered the water? No; the water was simply a road or path for the energy (created by the rock dropping into the water) to travel on. Sound traveling in air is very similar. Any vibration (or noise) will create a wave of sound, which transfers that energy over a highway of air particles. Once the energy makes it to the air particles in our ears, we hear the sound, just as we felt the water against our hand in the pool.

Figure 1.1
Water waves function in similar ways to sound waves. (Photo courtesy of iStockphoto, Deliormanli, Image #6752189.)

These sound waves traveling in the air are called *compression waves*. If you could see them, they would look very similar to waves in water except on a three-dimensional plane. That means that sound waves are traveling outward from a source in all directions—left, right, front, back, above, below, everywhere. When you think of sound waves, you might visualize them as two-dimen-

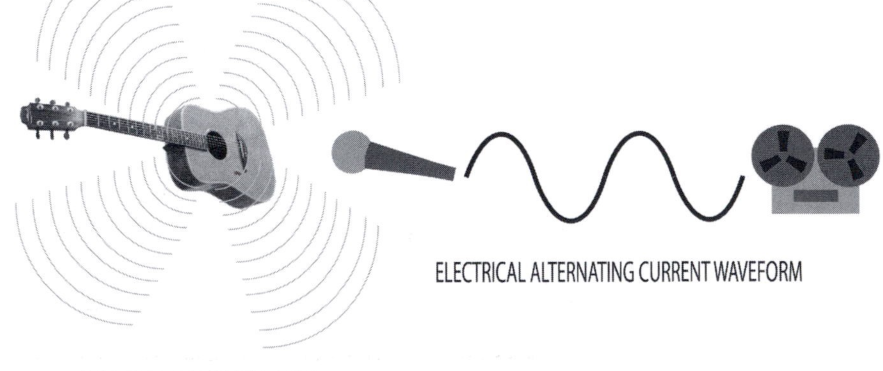

Figure 1.2
Compression waves are vibrations and variations in the air molecules around us. After being received by a microphone or some other transducer, these waves are an alternating electrical current and are represented as in the drawing. (Created by Ben Harris.)

sional waveforms, as they are usually drawn in recording programs and books. They look like this after they are captured by a microphone or pickup, but while traveling in the air in nature, they are three-dimensional compression waves.

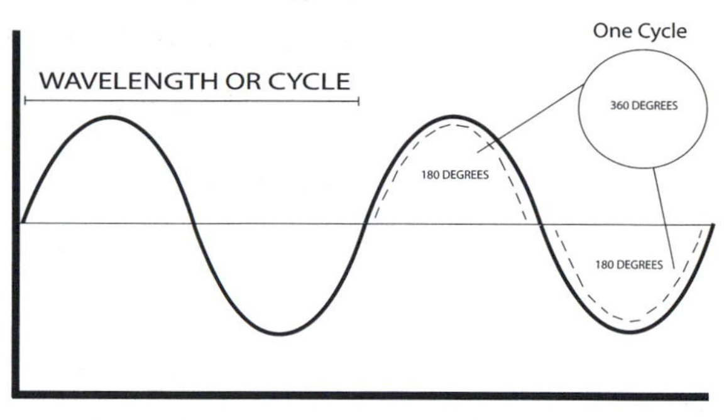

Figure 1.3
One wavelength or cycle is the time it takes for a sound wave to compress and undergo a rarefaction or go up and down and return to its original position. The frequency is how many times per second a waveform repeats. This defines *pitch* or *frequency*. (Created by Ben Harris.)

Frequency and Amplitude

There are two main measurable elements of a sound wave: frequency and amplitude. *Frequency* is the rate of forward and backward movement of a compression wave and up and down movement of a waveform. One wavelength is measured from the time that forward or upward movement (the compression) begins, then follows through backward or downward movement (the rarefaction) and returns to the start position. The amount of wavelengths or cycles that occur per second is the definition of frequency, and frequency defines pitch, such as 100 cycles per second, or 100 Hz.

Amplitude is the amount of variation between forward and backward movement, up and down movement, or compressions and rarefactions. Amplitude defines volume and sound pressure levels (SPLs). The higher the pressure, the louder the sound. These levels are defined by a logarithmic measurement called *decibels* (dB). Decibels are a relative measurement that, when applied to SPL, ranges from 0 dB SPL, the threshold of hearing; 60 dB SPL, the average level of speech; to 120 dB SPL, the threshold of pain. This measurement of dB SPL is different from dBs on faders of a console, but they both have an effect on volume and loudness.

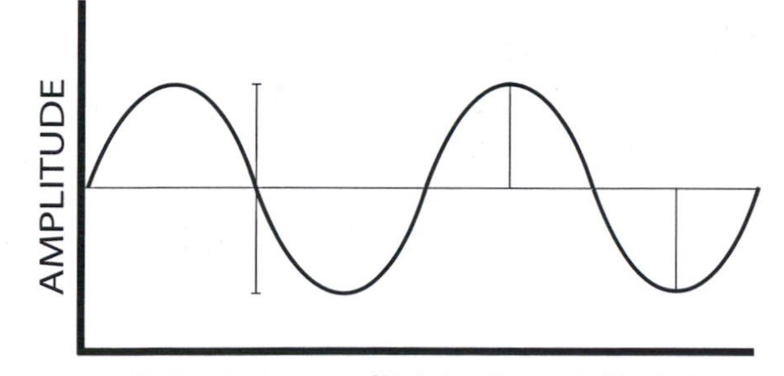

Figure 1.4
Amplitude is the variance of the up and down movement of a waveform. If it moves further up and down, there is more pressure and therefore a louder apparent sound. Amplitude helps define sound pressure level (SPL), volume, or loudness. (Created by Ben Harris.)

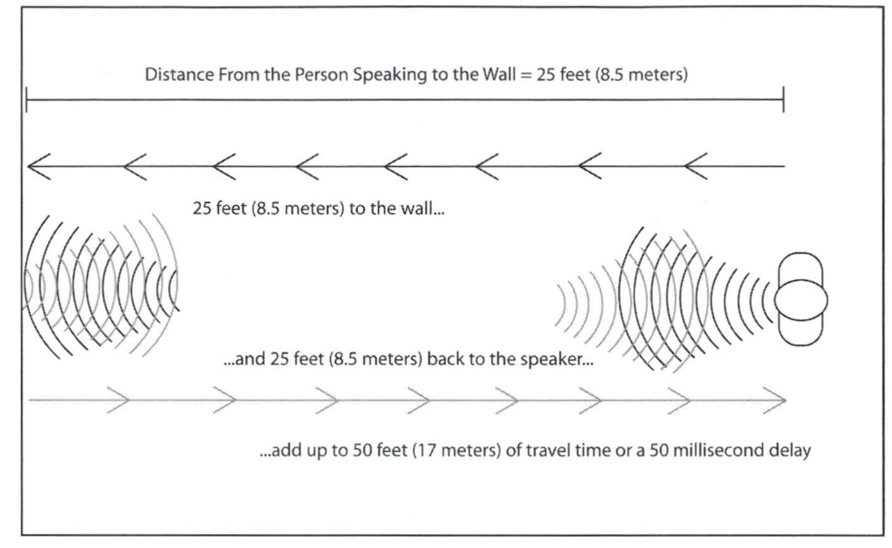

Figure 1.5
Delay off a reflected surface in a room can easily be calculated by measuring the distance to and from the surface. Each foot equals one millisecond, so if the sound travels 50 feet (17 meters), it will be delayed by 50 milliseconds when it returns to the speaker's position. (Created by Ben Harris.)

Delay and Reverberation

We mentioned earlier that sound waves travel in every direction, but what happens when they run into something hard? They don't stop; they bounce in another direction, and another, and another. It takes time for the waves to bounce here and there, so every time another bounce occurs, the sound gets more and more delayed. Sound travels at 344 meters, or roughly 1100 feet, per second. This means that if a sound travels 25 feet to a wall and then 25 feet back, by the time it gets back, it will be 50/1000ths of a second, or 50 milliseconds, delayed from the original. If the wave travels much farther than 50 feet, it will be delayed long enough to sound like an audible repeat. This is what happens in the mountains when you yell "Hello" and it sounds like someone is yelling back at you (an echo).

Reverberation (or *reverb*) is different from delay; reverb is the combination of thousands of different delays repeating at different times, coming at different levels, and coming from different directions. Reverb is the culmination of a sound bouncing off all the different surfaces of a room. Waves will keep bouncing around in a room until the energy has run out. So, the way to control acoustics is to either

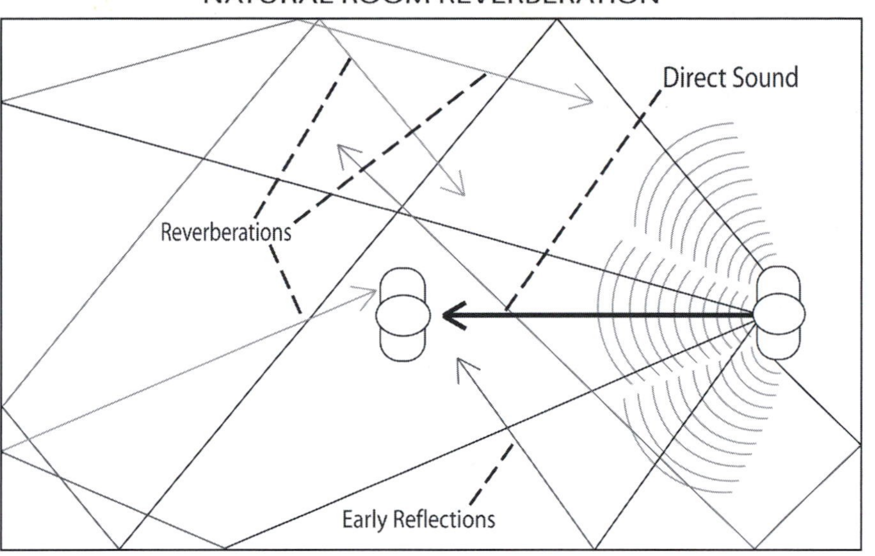

Figure 1.6
Reverb is made up of three elements: direct sound (the sound waves traveling straight from the speaker to the listener), early reflections (reflections bouncing off only one close surface), and reverberations (reflections that have bounced off multiple surfaces). (Created by Ben Harris.)

divert the energy with reflections or diffusion or stop the energy with trapping and absorption.

COMMON ACOUSTIC PROBLEMS

Why should you divert or stop acoustic energy if you don't even know what is causing the problem? There are a few main problems that occur in small rooms including flutter, comb filtering, and room modes.

Flutter and Comb Filtering

Flutter occurs when there are parallel hard surfaces and a sound wave simply gets stuck bouncing back and forth between the two. This sounds like a quick repeated delay and usually occurs in upper-mid and high frequencies.

Comb filtering occurs when two identical sound waves reach a microphone or ear at different times. If the delay between the two is very short, the two waves pile on top of each other and end up boosting some frequencies while canceling out others. This produces a hollow sound, which has a waveform that looks like a comb (see Figures 1.7 and 1.8).

Figure 1.7
Flutter echo occurs between hard parallel surfaces where the sound waves bounce back and forth between the surfaces, creating a quick ringing delay. (Created by Ben Harris.)

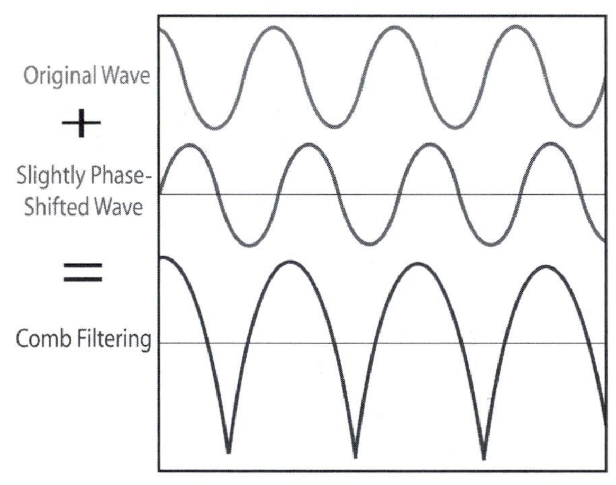

Figure 1.8
This image shows the effects of slightly delayed signals added on top of one another. It is called *comb filtering* because the waveform has the appearance of a comb. (Created by Ben Harris.)

Room Modes

Room modes (or *standing waves*) are frequencies that are boosted, diminished, or canceled based on the dimensions of a room. If the wavelength of a wave is the same length as the room, the wave will bounce back and pile on top of itself. In this circumstance it will cancel itself out. If the room is one-and-a-half times the length of the wavelength, the wave will double in amplitude or volume. This produces spots in the room where a frequency is boosted (nodes) or canceled out (antinodes), making the sound uneven and problematic.

Flutter and comb filtering are easy to fix, as we'll find out later, but room modes are definitely the most problematic and difficult to fix of the acoustic problems we've mentioned. Room modes are based on the dimensions of the room. For example, a room that is 10 feet by 10 feet will have a problem with 110 Hz because 10 feet is the wavelength of a 110 Hz wave. (If sound travels 1100 feet per second, it will travel 10 feet 110 times in a second.) This also means that there will be problems with 220 Hz and 55 Hz (twice and one half the wavelength). These frequencies are low-mid to low, which carry the muddiness in many instruments. Now your room is making your tracks more muddy because of the room modes.

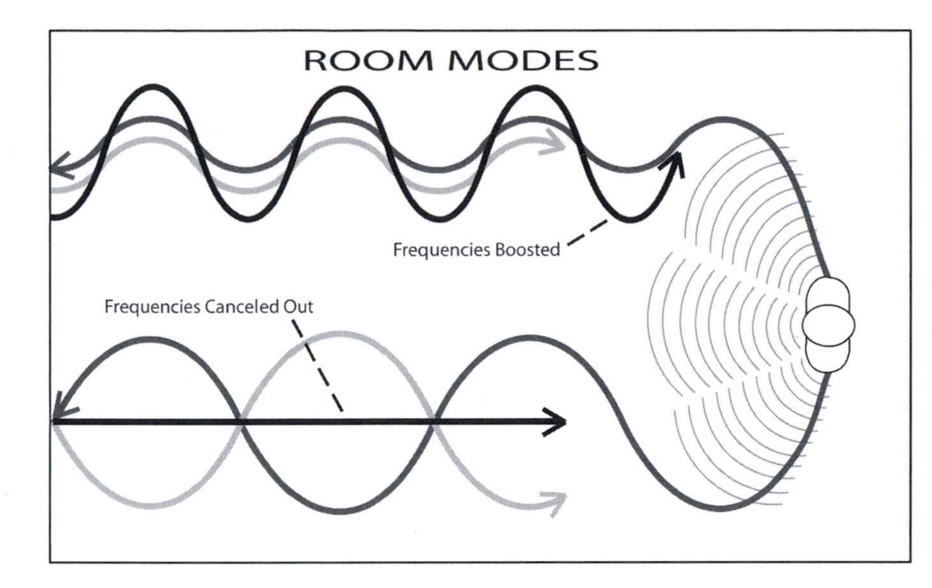

Figure 1.9
Room modes occur with the frequencies that have similar full, half, double, or any fraction of wavelengths of a room's dimensions. As we've seen, frequencies will either cancel themselves out or boost themselves, depending on what point of the cycle a wave is at when hitting a boundary and reflecting. (Created by Ben Harris.)

Figure 1.10
(Created by Ben Harris.)

Let's take another example, but this time we'll use a larger room. A room that is 30 feet by 30 feet will have a mode around 30 Hz. This is a very low frequency, almost beyond our audible range (20–20,000 Hz). If anything, this room will make your kick drum sound deep and awesome. To add to the problem, in small home studio rooms producing room modes in the muddy low-midrange frequencies (100–300 Hz), just putting foam on the wall does not solve the problem. Acoustic treatment solutions for room modes require mass and space to perform effectively against these frequencies. We will discuss some solutions to this daunting problem in a few chapters.

THE TWO TYPES OF ACOUSTIC TREATMENT

All acoustic treatment can be placed in one of two main categories. These categories are based on the goals, purpose, and function of the treatment. The two types of acoustic treatment are control and isolation. *Control* is treatment that is focused on getting rid of unwanted reflections and room modes and on balancing the frequency response of a room. *Isolation* is either keeping unwanted noise from getting into a room or keeping inside noise from getting out.

Control

Most of the problems discussed so far fall under the category of control. Flutter, comb filtering, and room modes can all be controlled by a combination of absorption, diffusion, and trapping.

Absorption

Absorption is basically achieved using fluffy stuff such as foam, insulation, or fabric. When a sound wave hits absorptive material, the energy of the wave is slowed down and disperses into heat in the material. Once the energy is stopped, the sound is finished. If the sound wave energy is not stopped, it will continue right through the absorption and keep reflecting until its energy has dissipated. Middle- and lower-range frequencies are usually slowed down but not completely stopped by basic absorption. These lower-frequency waves often have enough power or a long enough wavelength to pass directly through absorption without being slowed down at all. This is why basic absorption is generally effective for upper-mid and high-range frequencies, but not for low- and low-mid frequencies. Combinations of bass trapping and absorption are necessary to stop or significantly slow middle- to lower-range frequencies.

Bass Trapping

Trapping involves combining absorption with pockets of air so that the frequencies that pass through the absorption are trapped in the pockets of air. The size or distance of these pockets is very important. Based on the wavelength frequency chart, any space that is a fraction or multiple of a frequency length will have greater ability to reduce that frequency's energy. For example, if you place a 1-inch piece of absorptive material, it will mainly absorb 1000 Hz and higher. As soon as you place a 3-inch gap between the absorption and the wall, the lowest absorbed frequencies drop to 300–500 Hz. Bass trapping is a magical thing. The key is that the frequencies are permitted to pass through a surface but not return back through.

Diffusion

The problem with reflections is that they focus all the acoustic energy on one location. Absorption tries to stop the reflection, but *diffusion* simply scatters or redirects the energy. By scattering the energy, you keep it in the room while keeping it from becoming problematic. Diffusion is often used above and behind the listening position in the control room (where the mixing is done) and potentially every-

WAVELENGTH
CALCULATION CHART

Frequency (Cycles) (hz)	Speed of Sound (ft)	Wave-length (ft)	Wave-length (in)	Half Length (in)	Speed of Sound (m)	Wave-length (m)	Wave-length (cm)	Half Length (cm)
20	1129	56.45	677.4	338.7	344	17.2	1720	860
40	1129	28.23	338.7	169.4	344	8.6	860	430
50	1129	22.58	271	135.5	344	6.88	688	344
60	1129	18.82	225.8	112.9	344	5.733	573.3	286.7
80	1129	14.11	169.4	84.68	344	4.3	430	215
100	1129	11.29	135.5	67.74	344	3.44	344	172
150	1129	7.527	90.32	45.16	344	2.293	229.3	114.7
200	1129	5.645	67.74	33.87	344	1.72	172	86
250	1129	4.516	54.19	27.1	344	1.376	137.6	68.8
300	1129	3.763	45.16	22.58	344	1.147	114.7	57.33
400	1129	2.823	33.87	16.94	344	0.86	86	43
500	1129	2.258	27.1	13.55	344	0.688	68.8	34.4
600	1129	1.882	22.58	11.29	344	0.573	57.33	28.67
700	1129	1.613	19.35	9.677	344	0.491	49.14	24.57
800	1129	1.411	16.94	8.468	344	0.43	43	21.5
900	1129	1.254	15.05	7.527	344	0.382	38.22	19.11
1000	1129	1.129	13.55	6.774	344	0.344	34.4	17.2
1500	1129	0.753	9.032	4.516	344	0.229	22.93	11.47
2000	1129	0.565	6.774	3.387	344	0.172	17.2	8.6
2500	1129	0.452	5.419	2.71	344	0.138	13.76	6.88
3000	1129	0.376	4.516	2.258	344	0.115	11.47	5.733
3500	1129	0.323	3.871	1.935	344	0.098	9.829	4.914
4000	1129	0.282	3.387	1.694	344	0.086	8.6	4.3
4500	1129	0.251	3.011	1.505	344	0.076	7.644	3.822
5000	1129	0.226	2.71	1.355	344	0.069	6.88	3.44
5500	1129	0.205	2.463	1.232	344	0.063	6.255	3.127
6000	1129	0.188	2.258	1.129	344	0.057	5.733	2.867
6500	1129	0.174	2.084	1.042	344	0.053	5.292	2.646
7000	1129	0.161	1.935	0.968	344	0.049	4.914	2.457
7500	1129	0.151	1.806	0.903	344	0.046	4.587	2.293
8000	1129	0.141	1.694	0.847	344	0.043	4.3	2.15
8500	1129	0.133	1.594	0.797	344	0.04	4.047	2.024
9000	1129	0.125	1.505	0.753	344	0.038	3.822	1.911
9500	1129	0.119	1.426	0.713	344	0.036	3.621	1.811
10000	1129	0.113	1.355	0.677	344	0.034	3.44	1.72
11000	1129	0.103	1.232	0.616	344	0.031	3.127	1.564
12000	1129	0.094	1.129	0.565	344	0.029	2.867	1.433
13000	1129	0.087	1.042	0.521	344	0.026	2.646	1.323
14000	1129	0.081	0.968	0.484	344	0.025	2.457	1.229
15000	1129	0.075	0.903	0.452	344	0.023	2.293	1.147
16000	1129	0.071	0.847	0.423	344	0.022	2.15	1.075
17000	1129	0.066	0.797	0.398	344	0.02	2.024	1.012
18000	1129	0.063	0.753	0.376	344	0.019	1.911	0.956
19000	1129	0.059	0.713	0.357	344	0.018	1.811	0.905
20000	1129	0.056	0.677	0.339	344	0.017	1.72	0.86

Figure 1.11
Wavelength calculation chart.
(Created by Ben Harris.)

Figure 1.12
This image shows acoustic treatments placed a few inches away from the wall at a slant in order to have an effect on lower frequencies. Curved treatments are also visible; they provide both diffusion and absorption. (Photo courtesy of Wind Over the Earth, Ben Harris.)

where in the studio (where the instruments are played, sung, or spoken). Diffusion is the perfect fix for over-absorption because it keeps high-frequency energy in the room while still resolving issues such as flutter. A bookshelf can act as a natural diffuser, or a convex surface such as a column can be very effective as well.

Isolation

Isolation is the second main category of acoustic treatment. The purpose of isolation is to keep unwanted sound out, to keep sound from passing to one room from another, and to avoid annoying the neighbors. There is often confusion about the melding of control

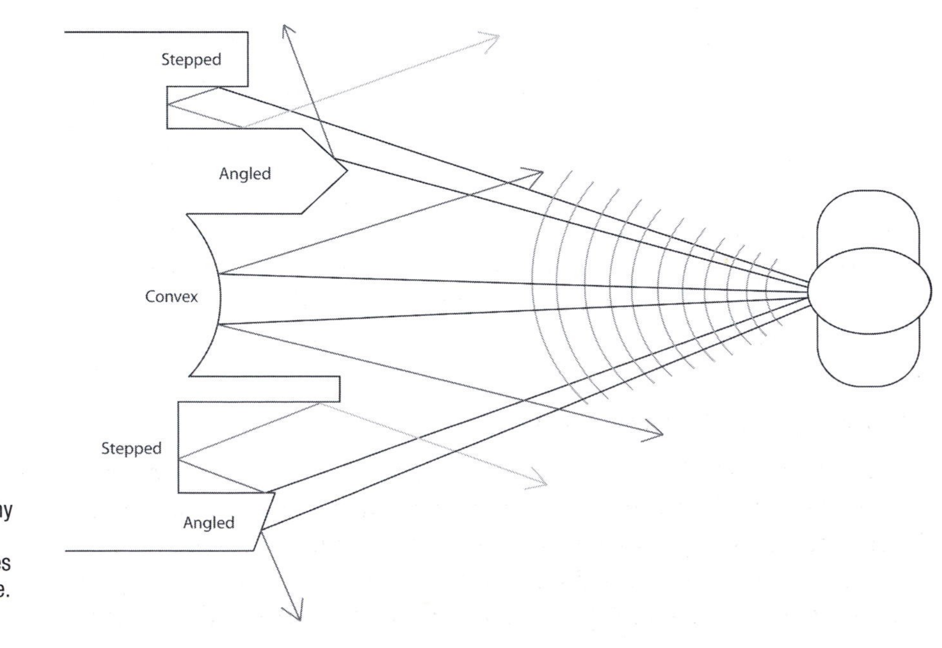

Figure 1.13
Diffusion can be accomplished with any surface that is not flat. Any combination of stepped, angled, or convex surfaces can create a diffuse plane. (Created by Ben Harris.)

and isolation, but efforts to do one will not usually help the other. The techniques for isolation in acoustics are not treatments on a wall; they are in the construction of the wall. The simple mantra, "mass … decoupling … mass …" is the key to effective isolation.

Figure 1.14
This image shows two windows (one with a slant) used to help isolate sound in a studio. (Photo courtesy of Sam McGuire.)

"Mass … Decoupling … Mass …"

What in the world does "mass … decoupling … mass …" mean? Basically, acoustic energy needs to be slowed and eventually stopped in order to isolate two spaces. If you just put up a bunch of mass (a thick wall), the acoustic energy will vibrate the wall and basically use it as a springboard to travel through to the other side. By placing a mass, a space of air, and another mass, you create a void where your acoustic energy gets stuck. The problem in real life is that it is impossible to completely decouple or separate two masses with only air. They have to touch each other somehow, the floor, the ground, or the like. Decoupling is accomplished in many ways, such as rubber washers between studs, foam padding, or nondrying caulk.

Room Within a Room

Many studios have room-within-a-room construction. This basically follows our "mass … decoupling … mass …" mantra by building a room and then placing another set of floor, ceiling, and walls inside of it. The key is that the inner room is decoupled from the outer room. If the floor can be sufficiently decoupled through layers of rubber, foam, or both, excellent isolation can be achieved. This could work perfectly if we didn't have to enter or exit the room, see out of it, or breathe. (See Chapter 3 for more construction techniques for acoustic isolation and room-within-a-room construction.)

CONTROL VS. ISOLATION

Control is the technique of using absorption, diffusion, and trapping to improve the internal acoustics of a room or space. Absorption, diffusion, and trapping each have a function, and the balance of all three is what makes an excellent acoustic space. No matter how good your acoustics to control your room, they will never fully perform the function of well-constructed isolation. Putting absorption up in a room may calm down the reflections and lessen the amount of sound traveling out of the room, but it will never stop the sound from traveling to adjacent rooms or structures. Construction techniques following the "mass … decoupling … mass …" mantra (discussed in more detail in Chapter 3) are the only way to isolate rooms and structures from each other and stop acoustic energy dead in its tracks.

CHAPTER 2
Planning

It is obvious that if a studio has acoustic problems, they need to be fixed, but many difficult questions go with solving these problems. These questions can be referred to as *programming* questions. Asking yourself the right questions and finding the honest answers can help make the most, acoustically, of any room or situation.

Figure 2.1
A good amount of thinking needs to go into planning what your studio will be used for. (Photo courtesy of iStockphoto, Joe Cicak, Image #5908297.)

PROGRAMMING

Programming is the process of finding out the best function for your studio, how to match your studio to your goals, and how to not waste time and money on things you don't need. The answers to these questions might not only apply for acoustic decisions. Many will help you make important decisions about equipment, comfort, and climate control. Some of the following are questions you might ask yourself before treating a room or building a studio.

Programming Questions

- *What will your studio be used for?* Voiceovers, composing, recording a band, and so on. The answer to this question will dictate size, the number of rooms, and the necessary equipment.
- *Does it need to be quiet all the time?* Will it be in use 24/7 (24 hours a day, 7 days a week), or will it be all right to do another

take because a train went by? If it is used 24/7, great expense will be needed for effective isolation and quiet heating, ventilation, and air conditioning.

- *How many musicians will be recording at the same time?* Will you be playing one instrument at a time, or will there be a full orchestra? This helps you decide the size and number of rooms, if rooms should be acoustically dead or more live, and how much equipment you might use in a given session.
- *Are you recording VO, music, sound effects, or the like?* Will a simple isolation booth do, or do you need a Foley pit or a large room for drums? This is another question to help you decide the number and size of rooms as well as necessary equipment.
- *What console are you using, if any?* A studio revolves around its recording console and needs to be designed accordingly. A large part of the acoustic design of a room is based around the console size, placement, and number of inputs and outputs.
- *Will there be one engineer working, or two, or four?* If there is more than one engineer, you will need sufficient space and an expanded sweet spot.
- *How big is your budget? Is it a $5,000, $20,000, or $100,000 room?* The answer to this question governs many of the decisions that need to be made.
- *How many clients will you have in your studio?* Will you be by yourself, have a producer, or a full band watching and listening? If you are working by yourself you will need less room, whereas if you have to accommodate a producer and/or band, you might consider a producer's desk, a larger control room, and a comfy couch.

FUNCTION

What is the function of your studio? Is it a one-room project studio for producing music, or is it a multiple-room, full-featured recording studio? Finding out the function is key to creating an efficient and effective recording studio. A lot of people putting together home studios without a vision of the function of their studio are unknowingly misguided into buying equipment that they might not really need. They might put a great amount of time, effort, and money into building two rooms when all they need is one big room. Let's look at a few scenarios of functions for studios and the acoustics, building, and equipment decisions that follow.

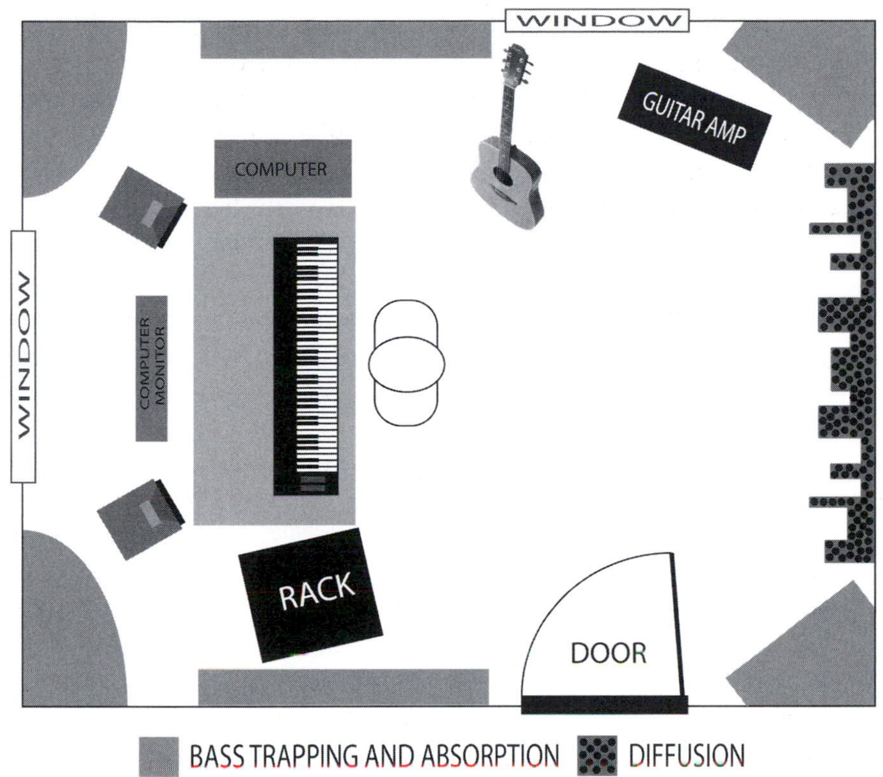

Figure 2.2
Floor plan of Acoustics Scenario 1 showing the basic layout of the equipment in the room and acoustics on the walls. The ceiling will need some acoustic treatment (absorption or diffusion) above the listening and recording position. (Created by Ben Harris.)

Acoustics Scenario 1 consists of a home project studio for a dentist who plays guitar and a little keyboard. He doesn't play in a band, but he likes to write music and wants to record it to share with his friends. He is only a hobbyist, so he will record whenever it is convenient. For construction he should build a one-room studio, although he might want to consider converting an existing closet into an isolation booth for his guitar cabinet. He doesn't need a big space to record drums, because he won't be recording drums. He will probably use loops or MIDI drums. He can record vocals and acoustic guitar in the same space where he mixes. He'll simply turn down the speakers and use headphones.

Acoustically he will need to focus on a balance of absorption, diffusion, and bass trapping. Absorption will probably be placed behind and to the sides of the speakers, diffusion will be put above and behind the listening/recording position, and he will have to focus specifically on bass trapping (most likely in the corners) because of the small room. Equipment-wise, he will need two quality channels of front end (see Chapter 5 for more details) for recording voice and guitar or both. This includes a high-quality stereo microphone preamp, one good vocal mic, one good acoustic guitar mic (potentially using the vocal mic for this purpose), and one good electric guitar amp mic (a good ribbon or dynamic mic). He will need only two stands, two mic cables, and at least one guitar cable. He will base his system around a powerful computer with an excellent recording/sequencing program, including a plentiful array of software synthesizers, a two-channel audio interface, and a MIDI controller keyboard. He won't need any external reverb, effects processors, compressors, or equalizers other than for his guitar rig. He will listen to his work on a small pair (5 or 6 inches in speaker diameter) of near-field studio monitors.

Acoustics Scenario 2 is a project studio for a regular performing band. They want to be able to work on new songs and upcoming albums at their own pace. The band consists of drums, bass, rhythm guitar, lead guitar, keyboard, and a lead vocalist. For construction

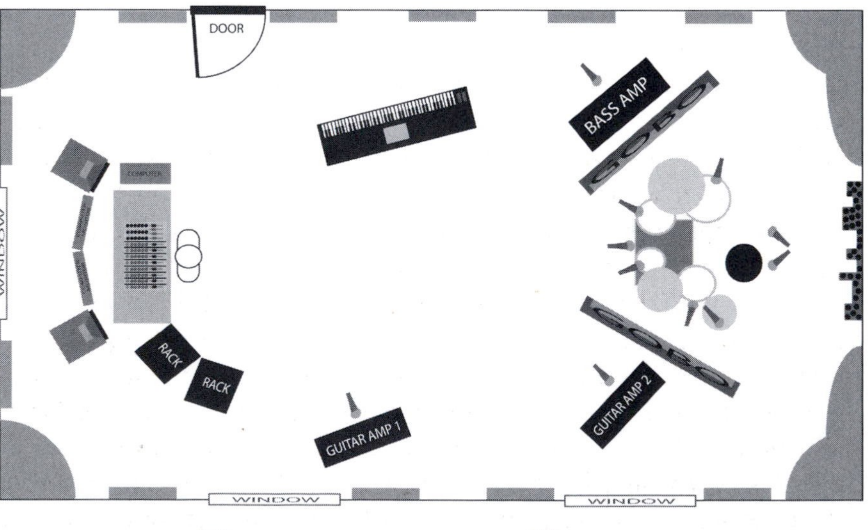

Figure 2.3
Floor plan of Acoustics Scenario 2, showing the basic layout of the equipment in the room and acoustics on the walls. The ceiling will need some acoustic treatment (absorption or diffusion) throughout the room. (Created by Ben Harris.)

they could either build one large room or do a three-room facility with a control room (for mixing), large studio (for tracking drums and large instruments or amps), and an isolation booth (for vocals or guitars). Any of these would work fine, with the three-room scenario giving better isolation and control but making communication more difficult and dramatically increasing the cost of construction. The one-room scenario would make isolation more difficult and monitoring more of a chore, but communication and production would be easier, with construction costs at a very reasonable level. For acoustics the one large room would be easier to treat with a balance of absorption, diffusion, and bass trapping, because there aren't as many bass issues in a larger room.

The one-room option would also require purchasing or making acoustic barriers, called *baffles* or *gobos* (as shown in the figure). These movable walls would help isolate the sound of the guitar amps from being picked up by the drum and vocal microphones. The three-room scenario would be more difficult since there are more isolation issues between rooms, requiring floating floors and multiple decoupled walls. The treatment for controlling the acoustics within the rooms would be a balance of absorption, diffusion, and bass trapping in the control room, a little more diffusion in the studio to make it more live, and more absorption in the isolation booth to make it more dead acoustically.

For equipment the band would need to focus on a front end capable of recording 12–16 inputs at a time. This would include microphones, stands, and cables for drums, bass, rhythm guitar, lead guitar, keyboard, and vocals. They would need 12–16 channels of high-quality microphone preamplifiers or eight channels of high-quality and eight channels mid-quality filler preamps. The recording system would be based around a decent-quality computer with a high-quality recording program filled with additional audio-processing plug-ins. A 16-channel audio interface would handle the analog-to-digital and digital-to-analog conversion needs of the system. The band might consider a 16-channel console to create multiple headphone monitor mixes or a personal monitoring system in which each member has control over his or her own headphone mix. External reverb, effects, compression, and equalization would not be necessary, because all processing would be done during mixing in the computer. A powerful pair of midfield studio monitors would be needed to provide sound for the entire band during playback.

ACOUSTICS SCENARIO 3

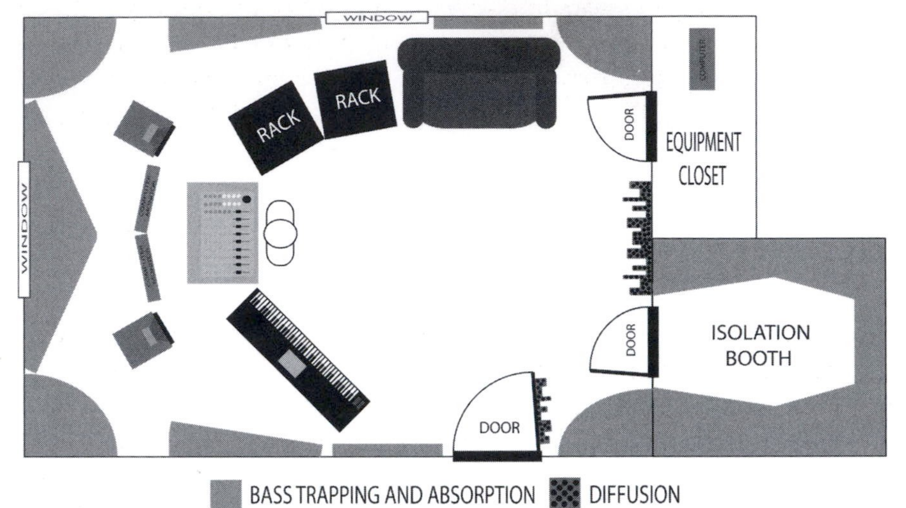

Figure 2.4
Floor plan of Acoustics Scenario 3, showing the basic layout of the equipment in the room and acoustics on the walls. The ceiling needs some acoustic treatment (absorption or diffusion) above the listening and recording positions. (Created by Ben Harris.)

Acoustics Scenario 3 is a producer's project studio. This producer wants to be able to start projects, do editing, and record overdubs in her studio. She plans on recording drums, large instruments, and ensembles in a separate professional facility. For construction she should build one room with an attached isolation booth. These should have a fair amount of isolation between each other and the outside world. The main room or control room would be used for recording, editing, MIDI production, and mixing. Acoustically this main room should be treated with a balance of absorption, diffusion, and bass trapping to provide a good recording, listening, and mixing environment. The isolation booth would have mainly absorption and bass trapping to be relatively acoustically dead.

The producer would want a small assortment of microphones and preamps in her collection, focusing around recording vocals, guitars, and other small miscellaneous instruments. The studio would be based around a powerful computer running one or more professional-level digital audio workstation (DAW) software programs. The system would be loaded with a large assortment of audio-processing plug-ins and software instruments. There would be at least one controller MIDI keyboard, other MIDI devices, and possibly classic keyboards and synthesizers. An eight-channel interface would be ideal to allow more options for recording more than one performer at a time.

Outboard equipment, besides microphone preamps, would be at the producer's discretion. She might like having a small collection of vintage EQ, compression, and reverb processors; or she might do all the processing in the box (in the computer) or at a bigger studio. Her monitoring system would include a high-quality pair of near-field/midfield studio monitors for detailed playback for tracking and mixing.

GOALS

The previous three examples clearly show how a studio can be focused around the goals of the owner. The basic premise is to clearly figure out what your goals are, then base your studio around meeting those goals. Achieving the primary goals is not usually the problem. The problem arises when the owner overachieves or sets unnecessary goals. For example, if you want to simply record songs for family and friends and only play guitar, don't build a drum room. If you will always play the dual role of engineer and musician, don't build a separate live room to record in; make your control room dual-purposed for recording and mixing. If you want to produce high-quality recordings on a budget, don't by a bunch of cheap equipment to accommodate four or five musicians recording simultaneously. Record your ensemble, piano, and drum tracks at a professional studio with a large room and nice equipment. Then bring the tracks home for overdubs (with your high-quality microphone and microphone preamplifier), editing, and mixing. If you plan on recording tracks in your studio and then taking them elsewhere to mix, don't spend your time and energy making an acoustically amazing mixing room; instead build a tracking studio (a studio focused on recording only).

Figure 2.5
How much money do you want to spend? (Photo courtesy of iStockphoto, Smartstock, Image #6510665.)

TIME AND MONEY

You might be wondering at this point why this chapter is about programming when we're supposed to be talking about acoustics. How do the function, goals, and money spent on a studio affect your studio's acoustic treatment? The answer is that programming has everything to do with it, and money is probably the biggest factor.

Most people overlook the importance of acoustics at first. They usually spend a little bit of money on their computer, software, and interface, a whole lot of money on a fancy microphone, and no money on acoustics. (Look to Chapter 5 to see why a fancy microphone only and nothing else fancy isn't always the best idea.) What most people don't realize is that no matter how nice the guitar, microphone, or preamp, if the acoustics are bad, everything will sound bad. That is the reason that a third of this book is about acoustics.

I can tell instantly if a track has been recorded in a poor acoustic space. Most home recordings sound muddy and muffled because the microphone was placed very close to the source to avoid the bad acoustics of the room (but this technique never works). Or sometimes people try to record with the microphone further away, then the track sounds like it has a permanent ringing nasty-bad acoustics reverb on it. Good acoustics make all the difference. If the acoustics are fair, you can back the microphone away from the instrument or vocalist and let the sound develop a little before it reaches the microphone. This usually gives the sound more body, life, and high-frequency response. I'm not saying that close micing is bad. If the acoustics are good, close micing doesn't sound as boomy, and you have the option to use an omnidirectional microphone up close. This gets rid of the boominess, maintains the presence of close micing, and lets some of the room acoustics in the track.

So, if you want your tracks to sound good (a goal), you should spend your resources (time and money) on getting a balance of good equipment and good acoustics (the function). If your resources are limited, you should not spend all of them on equipment and none on acoustics, or vice versa. Maybe you should get a little less pricey microphone, buy a nicer preamp, and spend some more time tweaking the acoustics of your room.

MAKING THE MOST OF YOUR ROOM

Most people building home studios are not constructing a room from scratch, tearing down walls, or dealing with spacious 12-foot-ceiling rooms. Most have only a small extra bedroom, unfinished basement, or garage to work with. The big question is, how do you make the most of your room or the space you have available? Do you use one room or two? Where do you put your desk and speaker in the room, and what are some tips to making your room work acoustically and functionally? This section will answer those questions.

One Room or Two?

Most people are under the impression that a studio should have two rooms and there should be a window between the two. This is what "real" studios look like on music videos, TV shows, and behind-the-scene featurettes, isn't it? This might be the case for many large studios, but many professional facilities use one room for everything, a vocal booth with no windows, or a camera and TV monitoring system. Even the studios with big mixing consoles record tracks in the control room at times. So, it's okay that you don't have to have two rooms with a window connecting them. You could have one large room and monitor through headphones during tracking, or you could have a separate room with a camera connected to a TV in your control room so that you can see the performer.

One Large Room

There are other reasons to have one large room instead of two, other than "you don't *have* to have two." Some of the best arguments have to do with acoustics. If you have a two-room studio, you don't want the drums from one room bleeding into the control room. The point is to have the rooms isolated from each other. Providing sufficient isolation between two rooms can be a difficult and costly undertaking. As mentioned before, you might want to build multiple walls, floating floors, and, of course, "mass … decoupling … mass." Building a window between the rooms is another costly endeavor. You could consider isolating the two rooms without a window and use the camera and TV scenario mentioned earlier.

Another acoustic benefit deals with controlling the acoustics within the room. An inherent problem with home studios is that they are located in small rooms, and small rooms have big low-frequency problems. As the room gets larger, low frequencies become less problematic and easier to deal with. If you build one large room instead of cutting it in half to make two small rooms, it will be easier to have a better-sounding mixing and recording acoustic space.

Desk and Speaker Orientation

When you are trying to decide where to place your desk and speakers, the biggest factors are probably ergonomics, sight lines, and

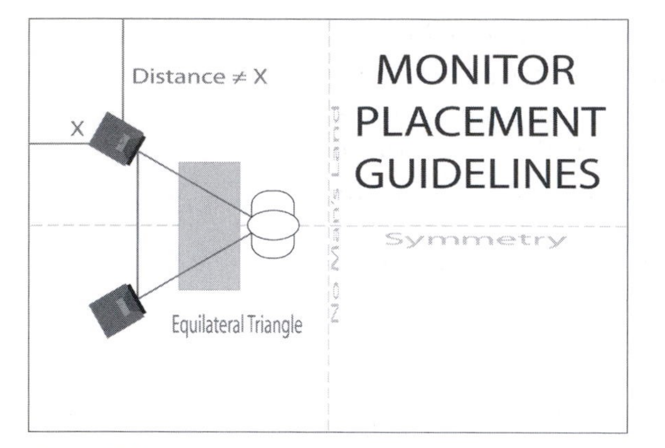

Figure 2.6
This image shows some of the basic guidelines for good monitor placement in a room. (Created by Ben Harris.)

convenience. These are all important, but acoustics should be weighing in toward the top. The placement of your desk and speakers in the room greatly affects how your mixing environment will sound. If your room is focused toward recording only, you can put your desk wherever you want. But if you want an acoustically balanced listening environment to help you create clean and accurate mixes, you need to pay attention to the following guidelines.

Symmetry

For the best stereo image, the desk and listening position should be centered between either side wall. By having the same distance between the left speaker and the left wall and the right speaker and the right wall, the stereo image will be balanced. If your image is not balanced, when you mix you will make it sound balanced in your room, but when played elsewhere it might seem right-heavy or left-heavy. This overcompensation is the reason that balanced acoustics is so important in a mixing environment.

If your room is too bassy, you will turn the bass down in your mix and your mixes will have a weak low end. If your room is unbalanced to the right, you will have mixes that seem left-heavy. If your room is symmetrical, each speaker will have a similar acoustic footprint in the room. This might not solve being too boomy or too bright, but your speakers will be balanced and you will have a better stereo image in your mixes.

Speaker Placement

If your desk is centered between the side walls in a symmetrical room, a lot of the trickiness of where to place your speakers is already solved. Simply place them equidistant from each wall and you're done.

But there is another factor. How far are your speakers from the front wall, or the wall that the back of the speakers face? The biggest rule to obey is this: Do not have the same distance between your speaker and the closest side wall and the speaker and the front wall (see above; Distance ≠ X). So basically, if your speakers are 3 feet from the closest side wall, they should not be 3 feet from the front wall.

They can be 2 feet, or 4 feet, or 1 foot, but not 3 feet. This is important because speakers not only emit sound out the front; they emit sound in every direction. That sound usually bounces and comes back toward the speaker. If the distance to the side wall and the front wall is the same from a speaker, those frequencies bouncing back to the speaker will come back simultaneously from each direction. This can cause some major buildup problems in the low frequencies in the room.

Closeness to the Front Wall

Many people might wonder why the previous rule is even an issue. They might say, "Don't you just put your speakers backed up to the wall anyway? What is this whole 1 foot or more thing about?" The distance behind a set of speakers is actually a huge issue acoustically. The sound emitting from the rear of the speaker that bounces off the front wall and now travels back toward the listening position blends with the sound emitting forward from the speaker and produces *comb filtering* (discussed earlier). The worst thing is that if your speakers are backed right up to the wall, you can get comb filtering, which destroys clarity, boosts some frequencies, and cancels out others. As you move the speakers away from the wall, the problems lessen. By the time you get a few feet from the front wall, the reflected signal is delayed enough to not cause any comb filtering or phasing issues.

At this point you would want to place some bass trapping and absorption behind the speakers to minimize the amount of sound reflected back toward the listening position.

Making Your Room Work

Basically your desk should be centered on one wall and your speakers should be set a couple of feet away from that wall, putting you toward the middle of the room, but not the complete middle (because you run into a whole other set of problems in "no man's land"). So, now that you are sitting close to the middle of the room, where do you have space to record? Good question. That is why all this information is a set of guidelines to factor in as you weigh your options considering acoustics, ergonomics, and convenience.

In some rooms the desk will have to be snug against the wall, but hopefully it will be centered on that wall, with some acoustic treatment behind the speakers in a symmetrical room. Maybe the solution is to have a desk on wheels that moves toward the wall during tracking and away during mixing. Maybe the space between the desk and the front wall is utilized as a storage area for guitar cases, boxes, or other studio equipment. I'm not going to give you scenarios for every possible room shape and tell you where the best placement of acoustics and equipment is for each of them. I'll simply lay down the principles that you should follow in making those decisions for yourself. So now you can look at your room and say, "I can utilize this space here while avoiding a bad acoustics reflection," and "I can put my desk on this wall because this gives my speakers better symmetry," or "I can record in this spot because even though it looks a little weird, it sounds better." Any room can work. The key is to follow sound acoustic principals to make it work for you as well as sound amazing.

Consult the companion website online at theDAWstudio.com for additional information and more acoustic scenarios.

CHAPTER 3
Construction

Acoustic treatment is not something that you can completely buy in a store and have it work perfectly. There is always some construction involved in making acoustic treatment work with your room. Most articles and books about acoustics drive you one of two ways: All acoustic treatment can be built by you, or all acoustic treatment must be purchased from brand-name acoustic manufacturers. I feel that the answer is often a combination of these two polarities.

BASIC CONSTRUCTION TIPS AND TRICKS

There are many times where you could easily build something that will work wonderfully for your specific acoustic problem, and there are other circumstances where something premanufactured, built to scientific specifications, or cut with lasers would be the most effective solution. In this chapter we discuss some common construction tips for homemade acoustic treatment materials as well as the strengths and weaknesses of available products in the marketplace. Keep up to date by consulting the companion website, theDAWstudio.com, for additional techniques, product reviews, and new technical advancements.

Bass Traps

First things first: A couch does not function as a bass trap. Yes, it does accomplish some absorption, and because it is big it does absorb lower frequencies than a thin piece of foam, but it is not a bass trap. A *bass trap* is a device that significantly absorbs low-frequency energy. This is usually accomplished by having sound pass through a permeable surface, move into a space of air, and then not return into the original space.

There are many bass traps that you can purchase that are very effective. Most of these fit into corners because these are places where

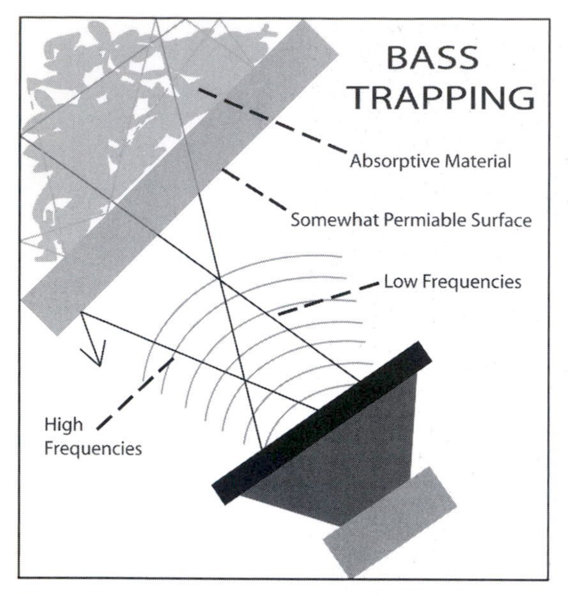

Figure 3.1
Bass traps let lower frequencies pass through a permeable surface, where the waves are then trapped and absorbed. Many bass traps reflect higher frequencies while letting low frequencies pass through. (Created by Ben Harris.)

Figure 3.2
A bass trap can be created by placing an acoustic panel diagonal in a corner. (Photo courtesy of Wind Over the Earth, Ben Harris.)

bass tends to build up (or load) and become problematic. There are also panel or membrane absorbers, which are considered bass traps, but use a different technique for trapping low frequencies. We'll talk about these later. The funny thing is that when it comes to traditional bass trapping (as mentioned first), custom-made devices and structures are usually the most effective and commonly used in professional facilities.

Building a Bass Trap

A bass trap can be made of a thin permeable wall with a space loosely filled with absorption behind it. Another type of bass trap, called a *circle trap* or *tube trap*, is basically chicken wire tied into a cylindrical shape covered in absorptive material and fabric with a stiff cover on the top and bottom.

The space in the middle is left empty to create a pocket of air to trap low frequencies. There are many alterations to this design using large PVC pipe with holes to pegboard covered in plastic. This type of trap has very simple construction but is often sold premanufactured at outlandish prices. The material and diameter of the tube have a direct relationship to which frequencies pass through, which frequencies are reflected, and which frequencies get trapped.

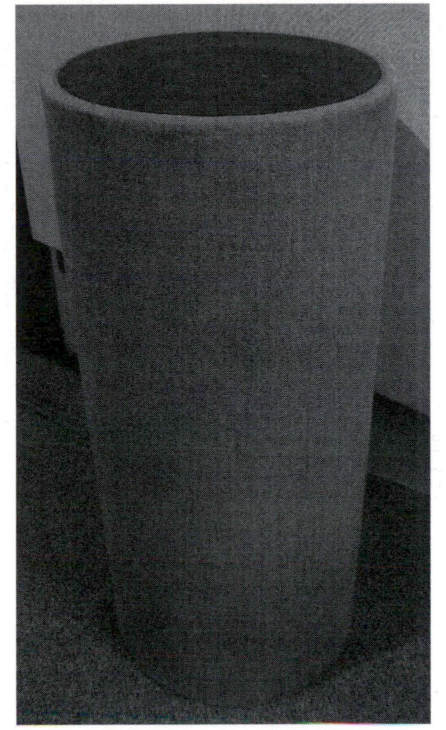

Figure 3.3
This is an example of a cylinder bass trap. (Photo courtesy of Wind Over the Earth, Ben Harris.)

A bass trap does not necessarily have to be a freestanding structure or something attached to a wall; it can be incorporated into the wall. You can build a second permeable wall next to an existing wall, cover it with fabric, and loosely fill the space with acoustic material such as insulation. The distance from the existing wall helps determine what frequencies are trapped, so it is a good idea to put the permeable wall on a slant, make it curved (convex), or make it stepped. This way it acts as broadband low-frequency absorption and trapping instead of focusing on a narrow frequency range. By making the outer surface somewhat hard and reflective while slanted, curved, or stepped, you can also create some diffuse surfaces.

Figure 3.4
A studio doesn't have to be in a square room. You can use slanted and stepped walls to help diffuse and scatter reflections. The walls are also an example of built-in bass trapping being built with stretched fabric over spaces filled with absorption. (Photo courtesy of Immersive Studios, Ben Harris.)

Diffusion

The main idea with diffusion is to scatter the acoustic energy back into the room to maintain the energy but not focus it back toward the listening position or a microphone. Three-dimensional diffusers are usually used above the listening position to avoid reflections off the ceiling from the speakers. These diffusers are used on the ceiling in the studio, especially over recording spots in the room, to avoid reflections from the source bouncing off the ceiling and hitting the microphone. Two-dimensional diffusers are usually placed behind the listening position and anywhere on the walls in the studio. Many people simply place diffusion behind the listening position because that is what everybody does, but it really doesn't work correctly unless there is a sufficient amount of diffusion in the back and enough reflection on the side walls to have that energy return to the listening position.

If that doesn't happen, the diffusion simply scatters the energy away from the listening position, which isn't bad, but it could be better. Diffusion in the studio is great when mixed intermittently with absorption. It doesn't work too well if there is only diffusion on one wall and absorption everywhere else. By placing diffusion, absorption, and reflection throughout the room, bad or unwanted reflections can be stopped and good reflections can be spread throughout the room to die out naturally. This will make the acoustics sound alive and natural.

You can use a three-dimensional diffuser on a wall, but energy that is diffused up and down is somewhat wasted. It is more than efficient

Figure 3.5
This image shows two-dimensional diffusion behind the listening position. (Photo courtesy of Immersive Studios, Ben Harris.)

Figure 3.6
These diffusers scatter sound on a three-dimensional field. They are most commonly used on the ceiling above the mix position in a control room or the tracking area in a studio. (Photo courtesy of Sam McGuire.)

to diffuse energy to either side on a wall. That is why a two-dimensional diffuser works well on walls. Ceilings need to diffuse energy in every direction, which is why three-dimensional diffusers are more common on the ceiling.

Figure 3.7
A simple bookshelf can provide a decent amount of diffusion in a studio. (Photo courtesy of iStockphoto, Luoman, Image #7106618.)

Buying or Building Diffusers

Many companies make expensive and very effective diffusers. RPG makes a two-dimensional diffuser (usually used on a wall) and a three-dimensional skyline diffuser (usually used on a ceiling). These diffusers sound great just about anywhere you place them. They are mathematically designed to equally scatter a range of frequencies. The frequency range of diffusion capable of a structure directly correlates to the distance of the gaps, spaces, or depth of the material. For example, if two slats of wood in a diffuser are 2 inches apart from each other, the lowest frequency affected will be 3500 Hz. (Refer to the frequency chart in Chapter 1.)

You don't necessarily have to buy or build a specific product to achieve more diffusion in your room. Any structure that has hard surfaces at different layers can achieve some amount of diffusion. For example, a bookshelf is much better than a straight wall. Just think of how different a room sounds when it is empty, as opposed to when it is filled with furniture, bookshelves, and wall hangings. All these elements add varying levels of absorption and diffusion.

There are lots of examples of three-dimensional art pieces that have multifaceted layers that are perfect for diverting acoustic energy while making a room look inviting and artistic. Be creative with what you place in your room, always keeping in mind how it might affect the acoustics. Any curved wall, pillar, or structure will scatter acoustic energy as it reflects it. The key is to have a convex surface such as a pillar (which scatters the energy), not a concave surface such as a bowl (which focuses the energy).

Figure 3.8
Acoustic foam is only effective in absorbing high-mid to high frequencies. (Photo courtesy of iStockphoto, Clay Cartwright, Image #116644.)

The Problem with Foam

Quick-fix acoustic foam products have infiltrated the home studio market in the last few years. These products are wonderful to market and sell for a few reasons. First, they are easy and cheap to make; second, they are easy for the customer to install (requiring only some glue or staples), and third, they appear to be a quick fix to a complicated problem.

But the problem with foam is that it isn't always a quick fix. There are often better and cheaper solutions, and though foam might fix some problems, it can't fix all of them. What often happens is that someone buys some foam (1, 2, or even 3 inches thick), places it all over the walls, and then proceeds to record or mix. They find that the foam effectively absorbs the high-mid to high frequencies, but it really doesn't do much for low-mids and lows (the real problem frequencies in a small room). It also fixes issues with flutter and ringiness. This is great, but now all the high end is gone and the low end is exactly the same as before, without any highs to balance it out. Now the room sounds insanely muddy and boomy, with no air or life to it at all. The problem with foam is that it leads users into a false sense of security, just like 50 SPF sunscreen and low-fat snacks. Foam does not fix everything; it can help, but it is not the end-all solution.

Figure 3.9
Fiberglass insulation such as this can be wrapped and covered with fabric or placed in bass traps to effectively absorb a wide range of frequencies. (Photo courtesy of iStockphoto, Branko Miokovic, Image #2534690.)

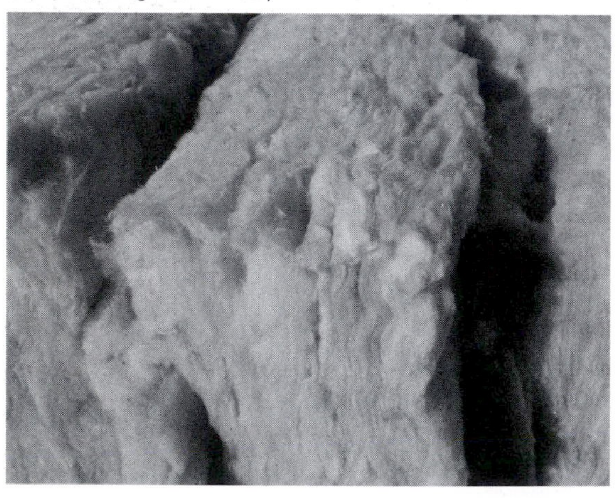

Alternatives to Foam

If you go to a professional studio and check out their acoustics, nine times out of 10 you will not see foam anywhere in the studio. This is because there are a handful of alternatives that are usually cheaper and more effective. The first is fiberglass insulation. This is used to insulate buildings and houses from heat and sound transmission. Pieces of this insulation are usually placed on walls, in frames, or in bass trapping structures and covered with fabric to prevent airborne fibers. This material has commonly been used for years due to its availability and effectiveness. It even comes in different sizes and with varying ratings that can help you decide which frequencies will be more effectively stopped.

The problem with this material is the fibers. These fibers rub onto your skin, go airborne and into your lungs, and some say cause adverse health affects to the user. This is why many people use alternate forms of insulation. There are many health- and eco-friendly products available for building materials, and therefore acoustic materials as well. Recycled denim and cotton insulation is a great example. This is similarly priced to fiberglass insulation but much easier to work with and just as effective. Some acoustic companies sell this same generic product specifically as acoustic treatment at twice the price. Recycled cotton and denim insulation can be used in all the same ways as fiberglass insulation except for the fact that it does not have to be completely sealed from contact with the air you breathe. This gives you limitless options as to how you utilize this insulation for acoustic treatment.

In addition, other types of insulation such as shredded newspaper, blankets, and carpeted wall coverings can serve as alternatives to foam for many acoustic absorption needs.

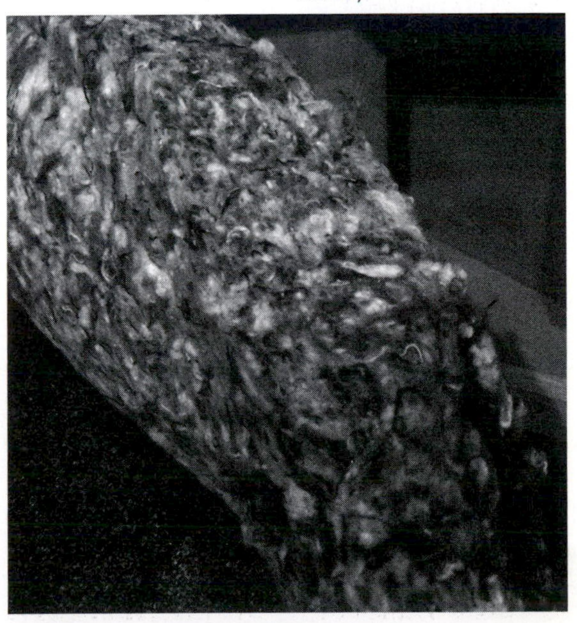

Figure 3.10
Insulation made from recycled cotton and denim has similar properties to fiberglass insulation and is much easier to work with. (Photo courtesy of Wind Over the Earth, Ben Harris.)

Building Additional Permeable Walls

By understanding the three types of acoustic treatment—absorption, diffusion, and bass trapping—and understanding how each is accomplished, you can easily build acoustic treatment that is multifunctional, ergonomic, and visually pleasing. Building additional permeable walls is an effective way of combining the three types of acoustic treatment techniques to effectively combat acoustic problems. Basically, building an additional permeable wall is building a large bass trap. By having a wall that sound can pass through with a space filled with absorptive material behind it, you can effectively trap a lot of troublesome bass frequencies. If you then curve, angle, or step that wall, you not only add diffusion, you vary the distance of the gap behind the wall. This variation directly corresponds to the frequencies that will be trapped. (See the frequency chart in Chapter 1.) Now your bass trap becomes more effective in a wider range of frequencies.

Acoustic ceiling tile, carpet, and pegboard are a few examples of material that can be used to create these permeable walls. The main thing to remember is that you are trying to create a balance of absorption, diffusion, and bass trapping. Cater your structure to the needs of the room and be prepared to change and rebuild if it doesn't sound right.

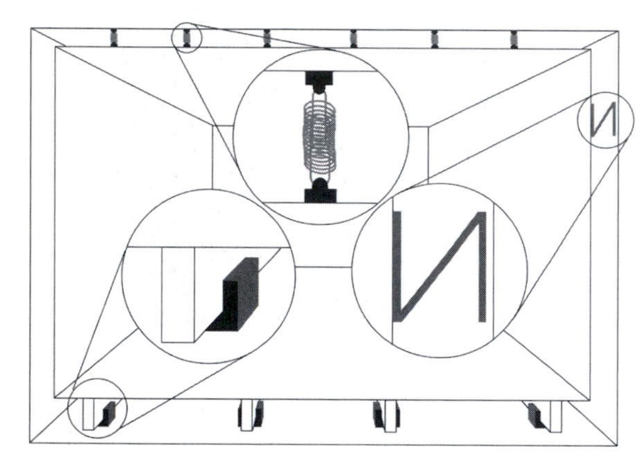

Figure 3.11
Here are three different products used in construction to help decouple walls from each other. These products are examples of many different techniques used to assist in acoustic isolation. (Created by Ben Harris.)

Wall and Floor Construction

While discussing isolation, the first chapter put a large focus on "mass … decoupling … mass," which essentially is building two structures while not letting them connect to each other. That's great; so we'll build a floor, put a layer of air, and another floor resting on the air. But … that's impossible? It is, and that's why there are many ways to essentially decouple structures while still obeying the laws of gravity.

There are some great devices and materials that make this possible, including springs, rubber, and foam. All these products help absorb vibrations, which is essentially decoupling the mass from the next mass or structure. If the two structures are connected, vibrations pass directly from one to the other. If there is a space of air between the structures, the vibrations will be trapped or greatly decreased because there is no mass to pass them along. If there is a pocket of air and then stiff brackets connecting the two masses to each other, the vibrations will pass through the brackets. The solution is to use materials such as springs, rubber, and foam to connect the structures but not rigidly couple (or connect) them to each other.

One way to decouple a raised floor from the main floor is to build the raised floor on studs resting on rubber pads. These pads will help decouple the raised floor from the main floor. Another technique is to use foam panels between the floors. This accomplishes the same thing as the rubber pads. Two walls (or ceilings) built next to each other can be decoupled by connecting them to each other with springs or a springlike *Z*-shaped bracket. These brackets will absorb vibrations and minimize them from passing from one wall to the other. If you have already effectively built a raised floor, build the walls on that floor and the ceiling on those walls so that the only connection that separates the room from the larger room are the rubber pads that the floor is sitting on. In larger studio design you can get even more crazy, creating two separate foundations separated by only dirt and then building each wall on an individual foundation. The vibrations stop when they hit the dirt and do not pass through to the other wall.

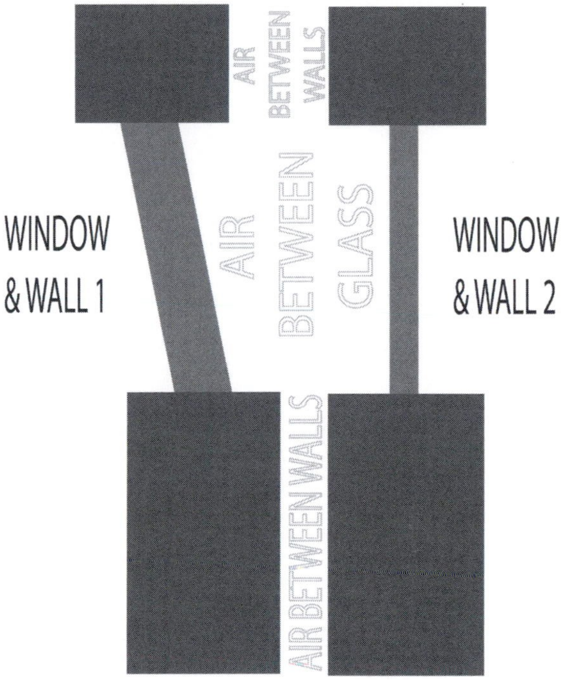

Check the DAWstudio.com for more details on acoustic products used for isolation in wall and floor construction.

Figure 3.12
A side view of dual window and wall construction utilizing varying glass thickness, one slanted piece, and air between for decoupling. (Created by Ben Harris.)

There are a few additional problems to avoid when placing windows in these walls so that you can see your clients. If both pieces of glass are parallel and the same thickness, they will vibrate in synchronization with each other, passing the vibrations right through to the other side. The solution is to use panes of glass of varying thicknesses so that they don't resonate at similar frequencies, then place one window on an angle to vary the distance between the two pieces, further preventing passing vibrations. The other big problem with windows (and doors) is that they are basically large holes and if air can leak through, so can sound. The solution is to seal all windows and doors as though you were sealing external windows and doors from the heat or cold outside. All these techniques will help isolate one room from another.

Panel or Membrane Absorber

Panel and membrane absorbers are a great solution to dealing with bass problems in situations with space limitations. These absorbers take up less space than traditional bass trapping because they function on a different acoustical premise. Instead of trapping the energy with absorptive materials and a pocket of air, panel and membrane absorbers utilize a thin layer of wood or heavy fabric that will vibrate but then dampen that vibration. Panel absorbers usually have a

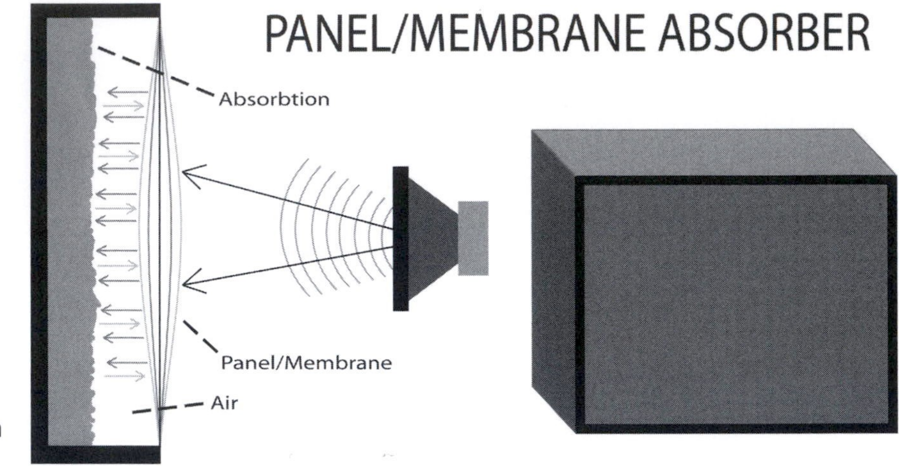

Figure 3.13
This top view of a panel/membrane absorber shows how acoustic energy becomes trapped. The front view shows that on the outside it looks like an ordinary box. (Created by Ben Harris.)

wood panel functioning as the absorber, and membrane absorbers can be built from roofing paper, synthetic leather, or any heavy tarp-like material. The idea is that the air behind the panel or membrane is sealed and contains foam or insulation (not touching the panel or membrane) so that when the front of the absorber begins to vibrate, the backside cannot keep up with the vibrations of the front. This is because the air on the backside is sealed and cannot move freely. As a result, the acoustic energy that excited the front of the panel or membrane becomes absorbed as heat into the air behind the panel and therefore does not return into the room.

These absorbers can easily be built with materials from a hardware store. A panel or membrane absorber is basically a shallow box with a thin layer of foam or insulation inside, pressed toward the back-side. There is a layer of air, and then the front is covered with either a panel or membrane made of the materials mentioned earlier. The depth of the box is in direct relation to the frequencies to be targeted. Frequencies that measure two and four times the length of the depth of the cabinet will be targeted the most by the absorber, but the material used for the panel or membrane is also a big factor to which frequencies are affected. The difference between using a wood panel or a membrane mainly affects the frequencies being targeted. Wood panels usually function more effectively on midrange and high frequencies due to their rigidity, whereas membranes usually function better on low-mid frequencies.

COMMON PROBLEMS AND SOLUTIONS

We've talked about a lot of common acoustic problems through the last few chapters, but the idea for this section is to address the issues not covered so far as well as revisit some covered earlier in greater detail. Some of the issues include room modes, equalization, and overabsorption.

Room Modes

Room modes are the frequencies that become problematic in a specific space based on the three main dimensions of a room; height, width, and length. Any one dimension in a room will produce problematic frequencies based on that dimension's length. For example, if the length of a room is 11 feet, then 100 Hz (one wavelength equal to the room length) will bounce back and forth on top of itself, effectively canceling itself out; 150 Hz (one-and-one-half wavelength) will pile on top of itself and double in amplitude; 200 Hz (two wavelengths) will do the same thing as 100 Hz.

This is problematic, but it becomes worse when we add two other dimensions to the room (height and width). What if your room is a perfect cube? Then each dimension has the same problematic frequencies, making the room an acoustic nightmare. Having one dimension twice the length of another is also problematic from the same reason as the cube. The way to calculate your room's modes is

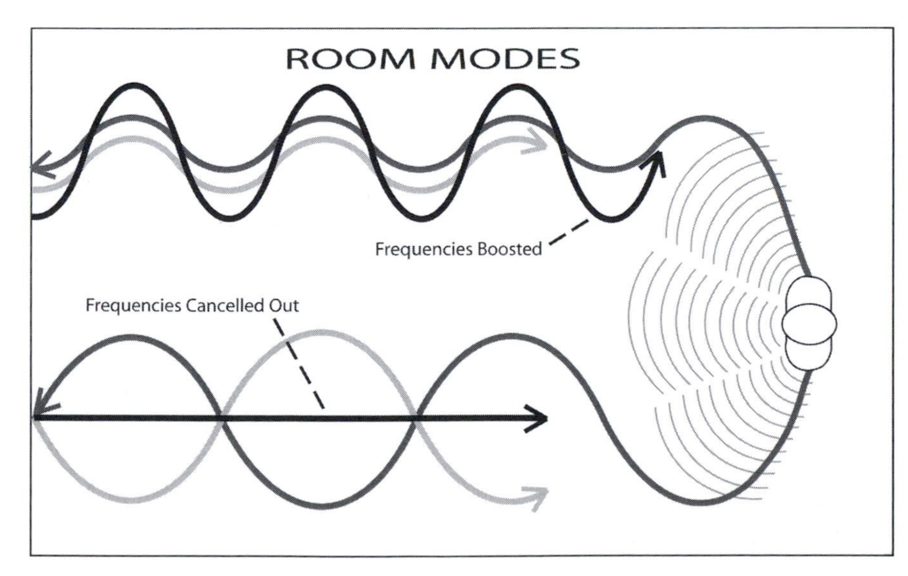

Figure 3.14
Room modes occur with the frequencies that have similar full, half, double, or any fraction wavelengths of a room's dimensions. As shown, frequencies will either cancel themselves out or boost themselves, depending on what point of the cycle a wave is at when hitting a boundary and reflecting. (Created by Ben Harris.)

to make a list of each dimension of your room and all the frequencies that have wavelengths that are multiples of your room's dimension (half the length, the same length, one-and-one-half the length, twice the length, and twice each of those). Once you do this for all three dimensions, find the frequencies that share two of the three dimensions, and those are your room modes. If there are common frequencies on all three lists, they will be specifically problematic. These are the frequencies that you should specifically target in you acoustic treatment efforts. Don't go too crazy; just focus on those specific frequencies. You want a balance of treatment with diffusion, absorption, broadband low-frequency absorption, and then specific treatment focusing on the problematic frequencies.

Equalization

Equalization is often misused in acoustics when it is used as the first line of defense. Often people will analyze a room, see the dips and bumps in the analysis, and go crazy with an equalizer trying to correct the problems. The problem with this lies in the previous paragraph. If you understand what is causing the frequency variations, you realize that equalization isn't going to fix the problem and will often make it worse. Think about it this way: If a room has a dip at 200 Hz because that is a frequency that cancels itself out as it bounces along the length of the room, adding more information at 200 Hz to the room will actually cancel itself out more and create a bigger dip. The same thing goes for bumps caused by frequencies bouncing on top of themselves in phase. There are also other factors that cause frequency deviation, such as reflections from the back of the speakers off the front wall that cause comb filtering, reflections off the side walls doing the same, and resonating furniture and items in the room. The problem is that trying to solve all these complex problems with a little equalization is simply illogical.

The logical approach to using equalization is to not use it as the first line of defense. First use techniques discussed in these chapters to correct these complex problems occurring in a room. Then, when you've done all that you can with physical acoustic treatment and it still isn't as flat as you would like, add a little equalization to balance it out. Don't just throw on a cheap three-band or anything. If you are going to equalize your room this way, you have to use a high-quality graphic or parametric equalizer. The adjustments are going to be very small—probably no greater than one or two dB of gain or attenuation (turning the signal down).

Absorption

If you go to a music store and tell the salesperson that you need some acoustic treatment for your studio, what will he suggest? Foam! But foam is a small and inadequate solution to a big and complex problem. The main reason that foam is readily available at music and recording equipment retailers is because it is an acoustic solution that can easily be packaged and sold. The problem is that it is not the end-all solution.

What happens is that people notice that their home-recording space sounds ringy (flutter echo) and boomy (low-frequency room modes). They buy a bunch of foam, glue it on their walls, and now the ringiness is gone but the boominess is worse. This occurs because the foam does not do much for the low and low-mid frequencies. It is great at getting rid of flutter echo and dampening the high and high-mid frequencies, but that doesn't do anything to solve the boominess. In fact, the boominess now seems worse because there isn't as much high-frequency information to counterbalance it.

What is the solution when you go back to the music store and explain the problem? Buy bass traps. Now you drop even more money on foam and glue these up in the corners of the room. Now it sounds a little better. The low-mids are not as horrible, but the low frequencies are as bad as ever and you still have no high frequencies in your room.

This process could go on forever, but the answer is that there is no kit or perfect solution that you can buy at a store, glue to your wall, and solve all your acoustic problems. Each room is unique and requires different treatment, care, and considerations. By understanding acoustic fundamentals, room interaction, and treatment techniques, you can individually treat a unique room to create an excellent recording and mixing environment.

Keep up to date with additional building tips and tricks, new building materials, and more detailed solutions to common acoustic problems at theDAWstudio.com.

SECTION 2
Equipment

CHAPTER 4

Computers and Recording

The home-recording revolution of the past 10 years has been possible because of one key element: the personal computer. In the 1970s and '80s home studios required a good-sized mixing console, a tape machine (reel or cassette), and a bunch of outboard processors (compressors, reverbs, and delays). In the 1990s the ADAT machine replaced the tape machine, making higher-quality recording available to the home studio user for a cheaper price tag. This was great, but the price tag was still around $5,000 to $10,000.

By the year 2000 computer processors were getting fast enough to handle basic recording and mixing tasks. In a couple of years the basic price of a powerful system was surprisingly affordable. Now for the price of a computer and an extra $1000, anyone can have a system that blows away the $10,000 home studios of 20 years ago.

DIGITAL AUDIO WORKSTATION (DAW)

A DAW, or digital audio workstation (pronounced *daugh* with a heavy southern drawl, or simply spelled out "D-A-W"), is a system

Figure 4.1
An example of a hardware DAW. It looks like a digital mixer, but it has a digital recorder built in. (Photo courtesy of Mike Simms, Ben Harris.)

designed to record, edit, and mix digital audio. There are two main types of digital audio workstation: standalone and software based. The standalone DAW is a contained hardware unit built with faders and an LCD screen. It has core internal software usually running on a small computer built into the hardware unit. Most of these have a built-in CD drive, four to 16 simultaneous inputs, and the ability to record and mix 16–48 total tracks. These units are wonderful for simple setups, band practice recording rigs, and songwriting scratch pads because they are very simple and easy to use. The problems with these units are that they are often difficult to use for advanced features (because of an LCD screen and lots of menus), you could quickly reach your track limitations, and they are not upgradeable. Okay, maybe you can upgrade by selling your unit for $1/10^{th}$ the price you paid for it and buying a brand-new one.

Software-Based DAW

The second type of digital audio workstation is a software-based DAW. These are computer programs that run on your personal computer. The main five software DAWs are Pro Tools by Digidesign,

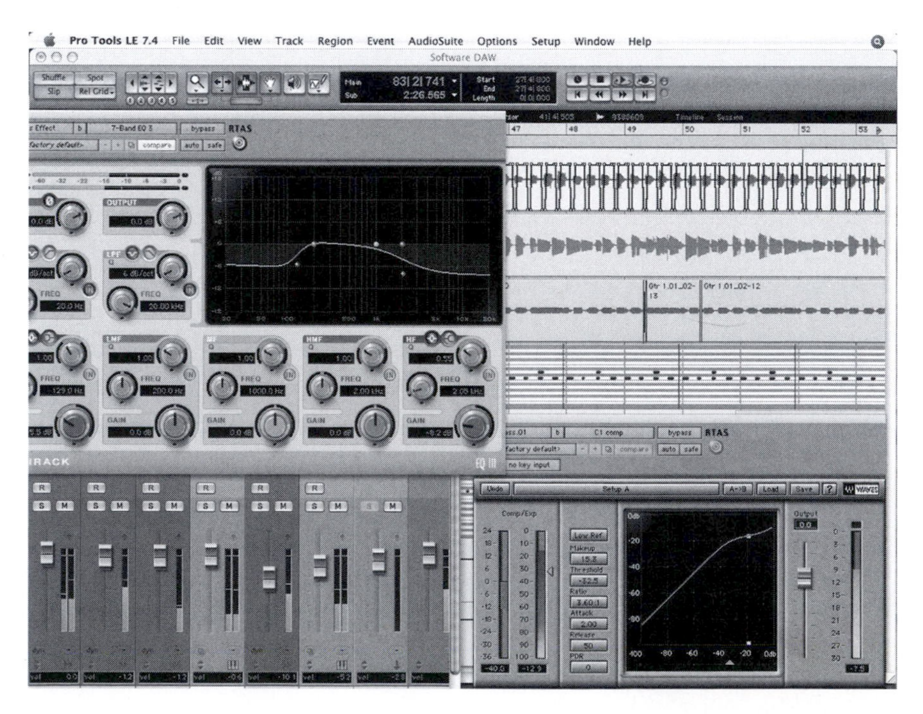

Figure 4.2
Pro Tools 7.4, a software DAW program displaying the mix and edit windows, an EQ plug-in, and a compressor plug-in from Waves.

Logic Pro by Apple, Cubase by Steinberg, Sonar by Cakewalk, and Digital Performer by Mark of the Unicorn (MOTU). There are a couple others vying to break into the top five positions, including Live by Ableton and Audition by Adobe. Each of these programs includes basic audio recording, editing, and mixing functions and much more.

There are four main types of music-based general software program functions that were once found in separate programs but now usually bundled together in most major software DAWs. These four types are digital audio recorder, MIDI sequencer, virtual instrument, and music notation editor. A *digital audio recorder* is simply a program that can record, edit, and mix digital audio information. A *MIDI sequencer* records, edits, and mixes MIDI information. A *virtual instrument* is a program that receives MIDI information and assigns it to different sounds. And a *music notation program* enables the user to put notes on a page and print a musical score. Most software DAWs incorporate all four of these types of programs in their software. The combination of all these functions is what makes software-based DAWs such a wonderful and useful tool.

A software-based DAW utilizes the power of a personal computer to run recording, editing, and mixing tasks. A system like this has three main components. First, the computer functions as the core power provider and processor of information, and it provides a place for audio data to be stored. Second, the DAW software provides the graphical user interface (GUI), the features to create music, and the ability to record and process audio. Third, the audio interface (or sound card, explained in Chapter 6) allows audio to be converted from analog to digital information and back, so it can be recorded to and played back from the computer. The beauty of a software DAW is that if one of these three things becomes outdated, all three do not have to be replaced. For example, you can easily get a new computer but continue to use your audio program and interface. Software manufacturers are always upgrading and updating their program software to keep it current. The user often outgrows hardware interfaces, or bigger and better interfaces become available. The majority of DAW programs are compatible with nearly any audio interface, from your computer's sound card to a $3000 two-channel unit. One program, Pro Tools, only functions with its own proprietary hardware interfaces.

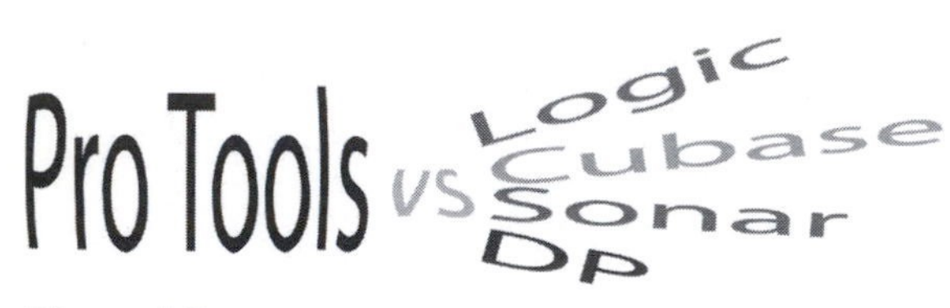

Figure 4.3
(Created by Ben Harris.)

Pro Tools vs. Everything Else

In the world of software DAWs, Pro Tools has the largest majority of users. There are many raging debates about why this is the case. The truth is that there is a distinct line between Pro Tools and all the other DAW software applications. The main difference is that Pro Tools started as a digital audio recorder and editor and over the years has added MIDI sequencing and virtual instruments. Other programs such as Logic and Cubase started as MIDI sequencers, later adding virtual instruments, music notation, and digital audio features. This division still persists and is best described as calling Logic, Cubase, Sonar, and Digital Performer *instruments* and Pro Tools a *machine*. Pro Tools is a machine because it is functional and allows you to easily do nonmusical or musical recording tasks without trying to assist in your performance. Other DAWs function as an instrument by incorporating built-in features that seamlessly assist in the creative and music-making process. To some people the music-making features get in the way. To others Pro Tools appears to be uncreative, dull, and boring. As you choose the DAW that's right for you, the real question to ask yourself is, do you want the program to create and perform with you, or do you want it to be functional and transparent while letting you do the creating and performing?

When it comes to professional recording facilities, Pro Tools is the standard. This is because it is simple to record, edit, and mix live instruments without other features getting in the way. It also has a standard set of key commands and simple two-window design. Other programs have multiple windows, with everything being customizable to function specifically for the one user. For that reason, these programs are more popular with composers who need a customized "instrument" to assist them in composing. Basically, if you want a program to simply record, edit, and mix live instruments with maybe some additional MIDI production, Pro Tools is the best option. If you plan on doing extensive MIDI and loop-based production and composition, one of the other programs might suit you best. There are countless other reasons to go one way or the other that are individual to each person, including which operating system is being used.

MAC VS. PC

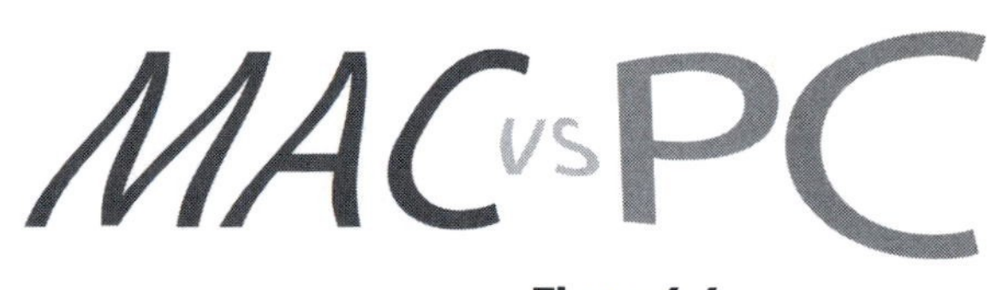

Figure 4.4
(Created by Ben Harris.)

This is a really sticky debate filled with misconceptions, lies, and biased opinions. Macintosh computers and their proprietary operating system are direct competitors to the open hardware market utilizing Microsoft's windows operating system. This competition formed when PC hardware and the Microsoft operating system became open source. This meant that any manufacturer could build hardware devices, software programs, or computer parts to use with the Microsoft operating system. This introduced competition, driving down the price while also bringing incompatibilities, low-quality products, and viruses. At the same time, Macintosh decided to continue building its own hardware and operating system to maintain high quality, security, and compatibility. Consequently, computers running Windows became less expensive, with more incompatibilities and more options, whereas Macintosh computers were more expensive and very compatible, with fewer options. The division continued as more word processing, spreadsheet, and business-based programs were developed for the Windows operating system, and music, multimedia, and artistic programs were developed for the Mac. Apple had a small, loyal following willing to pay higher prices for a stable, reliable, and high-powered system. Windows provided for everyone else, including the majority of the business world and home users. In recent years this division has become blurred as all types of programs have been developed for both platforms. When Macintosh adopted the use of Intel processors in 2006, the division blurred even further. The Intel processors forced Macintosh to redesign their operating system and hardware, making it possible to run Windows on a Mac.

The Legitimate Reasons to Choose

There is a small list of legitimate reasons to choose one operating system over the other (Figure 4.5). Any other reason not mentioned here is basically justification of a biased opinion. The list includes the following (see Figure 4.6):

- You are very comfortable with one operating system over the other and are not ready to learn a new feature set.
- You have chosen to use a program that either works better on one system or is only found on one.

Mac vs. PC Common Misconceptions

Misconception	The Truth
Macs are more expensive than PCs	First Apple only sells top-of-the-line processors and equipment in their current computers, while you can buy a brand new PC with pretty wimpy hardware. Second, there are many comparison charts (see the website) that show how in order to get equivalent hardware and software to a new Mac you have to pay the same price for a PC (even if you build it yourself).
PCs don't work for recording music	PCs are used to make music all over the world, especially in home studios. It is true that in professional recording facilities, less than 10% use PCs for recording, but they are using them and they are working.
Macs are not as upgradeable as PCs	This is true only in the mind of computer nerds, but for the average person Macs and PCs are equally easy to upgrade. The major hardware components that average people upgrade on their computers such as RAM, hard drives, optical drives, and video cards are identical to upgrade on Mac and PCs.
Freeware and shareware is only available on PC	There is an ever increasing amount of freeware and shareware programs available for Mac, but there are still more programs available for PC

Figure 4.5
(Created by Ben Harris.)

Mac vs. PC Comparison Chart

PC	Mac
PRO: PCs are the first choice for businesses creating a customized software/hardware solution for running their business.	PRO: Macs are still the first choice for most professional multimedia, video, and audio facilities.
PRO: PCs are the first choice for gamers and people who want to upgrade their processor and video card every six months.	CON: Macs still don't have a very great selection of games.
CON: PC computers cannot legally run any Mac operating system.	PRO: Mac computers can run both its own and a Windows operating system simultaneously or with a dual boot.
PRO: There are thousands of inexpensive hardware add-on options available.	CON: There is a minimal amount of hardware add-on options and many of them have no inexpensive option.
CON: PCs are susceptible to millions of viruses, constantly break, and require lots of maintenance.	PRO: Macs have zero documented viruses, rarely break down, and require practically no maintenance.
PRO: PC support and service is abundant and inexpensive.	CON: Mac support is hard to find and expensive.

Figure 4.6
(Created by Ben Harris.)

- You have family members who work for one of the companies in question.
- Your clients require that you run one system or the other.
- You are a gamer and have a deep desire to upgrade your computer hardware every few months.

COMPUTER HARDWARE

It's important to understand the main elements of a computer when you're configuring, optimizing, or maintaining a system. These elements include the processor, the motherboard RAM, USB and Firewire ports, hard drives, video cards, optical drives, sound cards, and PCI slots. Let's take a look at each.

The Processor

The processor is the heart of your computer. It is the brain, the place where all the calculations are made. It determines the power and speed of your system. Processors are usually what you see measured in gigahertz, such as a 3 GHz processor. Most processors are now dual or quad core, meaning that they are two or four processors synchronized and connected together to share the tasks of running the computer. This makes the processors very efficient, but many programs still don't take full advantage of the power available to them. You usually want to get the most powerful processor that you can afford.

Figure 4.7
A processor similar to what might be found at the heart of your computer. (Photo courtesy of iStockphoto, Murat Koc, Image #5887846.)

The Motherboard

The motherboard is where all the elements of the computer attach so that they can be connected together. The motherboard usually determines how large a processor can be used and how much RAM

Figure 4.8
The motherboard, the place where all the internal components of a computer connect to each other. (Photo courtesy of iStockphoto, Jaroslaw Wojcik, Image #2962275.)

can be installed. The motherboard and the processor are basically the core of the computer. If you change these, you are basically getting a new computer.

RAM

RAM stands for *random access memory*. RAM used to be measured in megabytes (128 MB, 512 MB) but is now more commonly found in gigabyte sticks (1 GB, 2 GB). All motherboards have a maximum number of RAM slots and therefore a maximum total amount of RAM that can be installed.

RAM serves many purposes in the day-to-day activity of a computer. When programs open, their files are loaded into RAM so that the computer can quickly access information to run the program and therefore make the program run faster and smoother. Since programs run from RAM, having more RAM lets you run more programs simultaneously. The other main purpose of RAM is to function as a buffer.

Figure 4.9
A RAM DIMM carrying memory chips used to buffer information as it travels through your computer. (Photo courtesy of iStockphoto, Vasko Miokovic, Image #7086551.)

RAM buffers information (or creates a steady and even flow) between the software and the processor, between the hardware and the processor, and between hard drives and the processor. There is also internal RAM on many devices, including video cards, hard drives, sound cards, and printers. Audio programs use RAM to buffer incoming audio streams, load up sound files for virtual instruments, and route audio to the processor for plug-ins. The more RAM you have, the better. All audio programs have minimum RAM requirements, but it is usually good to have double or triple that amount.

USB and Firewire

USB and Firewire are connection protocols found on motherboards and are used to connect to external devices. USB, short for *Universal Serial Bus*, utilizes a serial transfer of information (one piece of information after the other). USB 2 has a transfer rate of up to 480 mbps

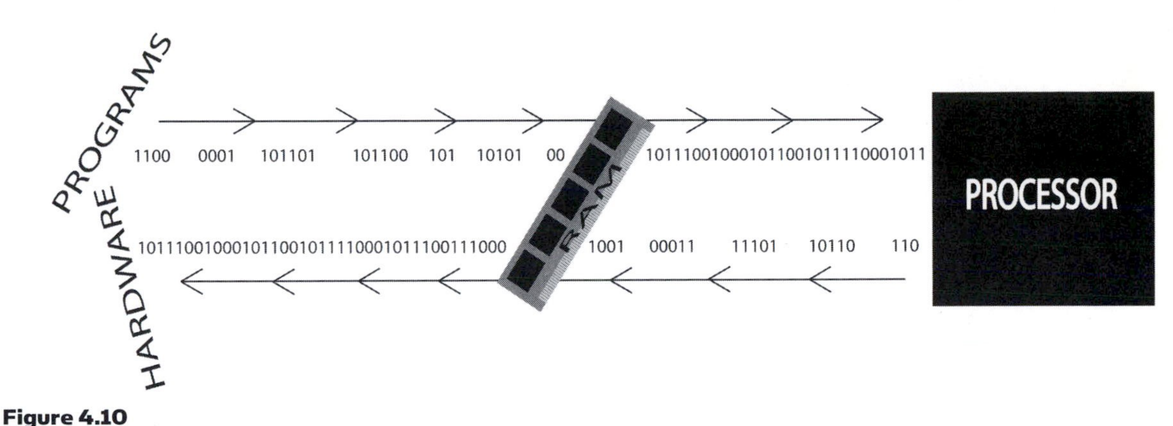

Figure 4.10
How RAM buffers information as it travels to and from the processor to and from programs and hardware. (Created by Ben Harris.)

(megabits per second). USB 2 is the standard connection for keyboards, mice, printers, still cameras, and two-channel audio interfaces. Firewire (IEEE 1394) utilizes a parallel transfer of information (multiple pieces are put into a packet and then unpackaged on the other side). Firewire 400 has a transfer rate of 400 mbps.

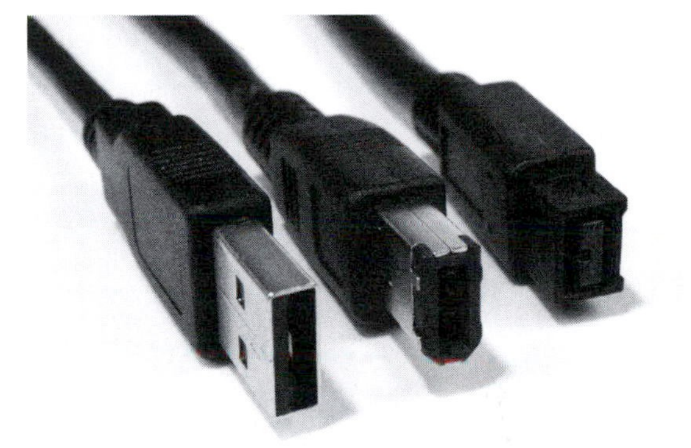

Figure 4.11
USB, Firewire 400, and Firewire 800 connections, from left to right. (Photo courtesy of iStockphoto, Dieter Spears, Image #1654385.)

Many people have the misconception that USB 2 is faster than Firewire when they compare 480 mbps to 400 mbps, but USB 2 runs *up to* 480 mbps (usually consistently averaging around 300 mbps), whereas Firewire runs *at* a consistent 400 mbps. Therefore Firewire is consistently faster. This is why it is the standard for video transfer and audio interfaces with more than two channels of input and output. Firewire also comes in a version that runs at 800 mbps, called Firewire 800, utilizing a different connector. Both Firewire and USB 2 can run up to 15 feet without amplification. To run farther, they need to be connected to a powered hub and therefore amplified.

Hard Drives

Hard drives are devices on which information is stored for an indefinite amount of time. These devices are sometimes referred to as *memory*, but this can be confusing, with RAM more commonly being

Figure 4.12
An internal hard drive with its protective cover removed to show the magnetic platter where data is stored. (Photo courtesy of iStockphoto, Brandon Laufenberg, Image #6693540.)

called memory. Hard drives come in two main forms: internal and external. Internal hard drives are placed inside the computer and connected directly to the motherboard. External hard drives are enclosures with USB or Firewire ports via which they connect to a computer. These are plug-and-play devices, meaning that they can be connected and disconnected without having to turn off the computer. They can be used to back up information or give additional storage space if a computer's internal drives are nearly full. Hard drives are somewhat delicate. They tend to break when dropped, run poorly when over 75% full, and sometimes simply fail.

Most recording software manufacturers list which external hard drives are good for playing and recording audio files with their program. Hard drives are one of the main factors contributing to a system's level of performance. Many errors can occur because a user is recording to a system drive, a drive that is too slow or fragmented, or a drive that does not have sufficient bandwidth to transfer information quickly. One suggestion from almost every software DAW manufacturer is to not record audio onto your system drive (the drive that contains your operating system and program files). The problem is that your system drive is constantly being accessed to find data for running your operating system and open programs. By simultaneously reading or writing audio files to the system drive, you are making that drive work harder than it needs to. By having a separate dedicated drive for audio, you can ensure that your system will run smoothly.

Drives are often defined by their *seek time*. This is the time from when a drive receives a command to recall a file until the file is actually retrieved. A faster seek time means that the drive will be able to play back and record all your audio files without any glitches.

Drives become fragmented because information is not placed on the drive in a linear fashion. Once you delete and rerecord information on a drive multiple times, little bits and pieces of files are placed in whatever spot is available. A *fragmented* drive is a drive that has this problem to the degree that it takes much longer to retrieve a file because the hard drive has to find multiple fragments of the file all over the drive to piece it together.

There are multiple connection formats for hard drives. Internal drives are now most often SATA drives, which can have a transfer speed of up to 3 Gigabits per second. This is much faster than the external formats mentioned earlier, but it's not as convenient. If you do have room for an additional drive in your computer, I would highly suggest you get one.

Hard drives are also used for a backup in recording. Drives can be finicky and can sometimes simply die. If they're filled with hours of work, this can be a disaster. Sometimes the information on drives can be recovered, but this can be expensive. I suggest having an additional external drive on which you back up any recording that you do. This way, if one drive dies, you still have the data on your backup. Once your recording is complete, you might want to put it onto a storage device such as tape backup (small digital data tapes), CD or DVD, or a hard drive that you put on the shelf.

Video Cards

Video cards hold the connection to computer monitors. As with anything else, you can pay more money to get higher-quality images, quicker response times, and multiple monitor connections. Video cards are a big deal with gamers and video editors, but for audio they just need to work and give you dual monitors if desired.

Two main types of video connectors are found on video cards. The *VGA connector* is the older analog format usually seen with the large bulky CRT tube monitors (although this is not always the case). The *DVI connector* is the new digital format that is much clearer and crisper and usually seen with sleek flat-screen monitors. The biggest thing to remember is that the DVI connector carries a VGA signal, and a $15 adapter can easily make that change. It is a lot more difficult to convert from VGA to DVI, so make sure that your monitors are compatible with the video card in a system.

Figure 4.13
A video card with VGA, S-video, and DVI connectors, from left to right. (Photo courtesy of iStockphoto, Jamie Otterstetter, Image #3710863.)

One concern that can affect performance is if you have a video card with insufficient RAM, it will use your system RAM when its RAM runs out. This can cause your system to

Figure 4.14
An optical drive that could be in a computer or an external enclosure. (Photo courtesy of iStockphoto, Michal Koziarski, Image #3189282.)

become bogged down. The solution is to have a video card with sufficient amounts of internal RAM.

Optical Drives

An *optical drive* is a device that reads and writes to optical media (CDs and DVDs). These drives come in internal or external options, similar to hard drives. There is an ever-increasing number of optical media formats available. First is the compact disc (CD), which is the format for audio CDs and data installation CDs. CDs can hold 74 to 80 minutes of music or 700 to 800 megabytes of data. Second is the digital video disc (DVD), which comes in two versions: single layer and double layer. Double-layer DVDs write information on two separate layers of the disc (which is why movies on DVD have a little hangup half way through the movie; the system is switching to the second layer). The two layers are on top of each other, with the upper layer being somewhat transparent so that the laser can read through it to the lower layer. A single-layer DVD can hold up to 4.7 gigabytes of information, and a dual-layer disc can hold up to 8.5 Gb.

Blu-Ray is the next step up, playing high-resolution video and allowing an enormous amount of data storage space from the same size disc as CD and DVD. This is made possible because it uses a blue laser instead of a red laser (hence the name). A single-layer Blu-Ray DVD is capable of storing up to 25 gigabytes of information; a dual-layer disc can store up to 50 Gb. All these formats are found on the same size disc, but not all players and burners will record all the formats. The nice thing is that each format builds on the other, so a DVD burner can burn CDs and DVDs and a Blu-Ray burn can burn all three formats. Not all burners can do both single layer and double layer, and the discs become increasingly expensive relative to their capacity. For audio applications you'll need a CD burner, but DVD and Blu-Ray burning capability can also be useful for data backup.

Sound Cards and Audio Interfaces

A *sound card* is a device in a computer that converts an analog audio signal into digital information, and vice versa, to enable recording and playback of audio. Almost every computer comes with some form of a sound card, either in a PCI slot or connected to the motherboard (as shown), to enable the user to play audio from a computer. Most consumer sound cards are not suited for use with recording applications. My general rule is that if you can buy it at a computer store, it is not high enough quality to use for recording audio.

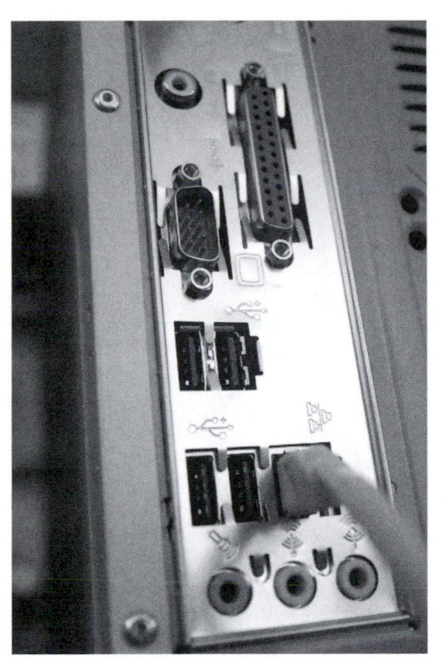

Figure 4.15
Three connections of a computer's sound card: a line input, a line output, and a microphone input. (Photo courtesy of iStockphoto, Roro Fernandez, Image #261428.)

The other rule is that the recording world uses audio interfaces, not sound cards. Sometimes these two look identical, but you buy audio interfaces from music and recording equipment stores and you buy sound cards from computer stores. Some computer stores sell audio interfaces, but they are usually the cheapest ones available. So, the sound card isn't that big a deal, but the audio interface is huge (as discussed in Chapter 6).

PCI Slots

PCI slots are ways of connecting additional devices or ports directly to a computer's motherboard. These slots are found at the back of desktop computers. Laptops usually have one expansion port, similar to a PCI slot. PCI slots currently come in three formats: PCI, PCIX, and PCIe. PCI is the oldest and slowest of the three. PCIX is basically an expanded PCI format. It uses the same bus architecture but is quite a bit faster. PCIX is backward compatible with PCI, but PCI is not forward compatible with PCIX. Finally there is PCIe (PCI Express), which is a completely different architecture than PCI. It is faster as well with a smaller connection. PCIe is becoming the standard for most PCI cards but is taking a while because the old

Figure 4.16
A standard PCI card. These cards vary among PCI, PCIX, and PCIe, depending on the position of the gold colored pins (shown on the lower part of the card). Be sure to get cards that are compatible with the connections on your motherboard. (Photo courtesy of iStockphoto, Konstantin Sukhinin, Image #1355568.)

cards have to be redesigned for this new format. There are some computers that come with all three formats; more and more come with just PCIe.

PLUG-INS

Plug-ins are found in many computer software programs and are basically small add-on programs that run within another host program. In the audio and recording world there are two main types of plug-in. One type is a virtual processor such as an equalizer, compressor, or reverb unit. These plug-ins are similar to buying additional hardware units for an analog recording rig. These processors can be placed in a DAW's software mixer to route audio signals through them. The second type is a virtual instrument plug-in. These plug-ins do not process audio, but they take MIDI information and assign them to instrument sounds. These are similar to synthesizers or keyboards that let you play a saxophone or drums using piano keys.

Stock vs. Third-Party Plug-Ins

Most software DAWs come with a collection of virtual processor and virtual instrument plug-ins. Some DAWs have a better stock collection than others, but all applications are compatible with third-party or additional plug-ins. Stock virtual processor plug-ins usually do the job and function well as equalizers, compressors, and reverb units, but they are mostly just getting by. They usually don't sound completely amazing. This is why there are third-party plug-ins. People want better-sounding options than what comes standard with the DAW. Some of the biggest virtual processor plug-in manufacturers include McDSP, Sound Toys, and Waves. Third-party virtual processor plug-ins are sold individually or in bundles and range in price from $100 for a single plug-in to $5000 for a huge bundle. As far as virtual processor plug-ins, I suggest utilizing the stock plug-ins first and then later looking into third-party options.

Stock virtual instruments vary greatly based on the application they come with. Most DAWs come with at least one general-purpose virtual instrument that usually includes a basic set of sounds, including pianos, guitars, brass, woodwinds, strings, and percussion. Most third-party virtual instruments focus on one family or type of instrument. For example, there are virtual instruments dedicated to just piano sounds, just drums, or just strings. These instruments can be

EXAMPLES OF STOCK PLUG-INS

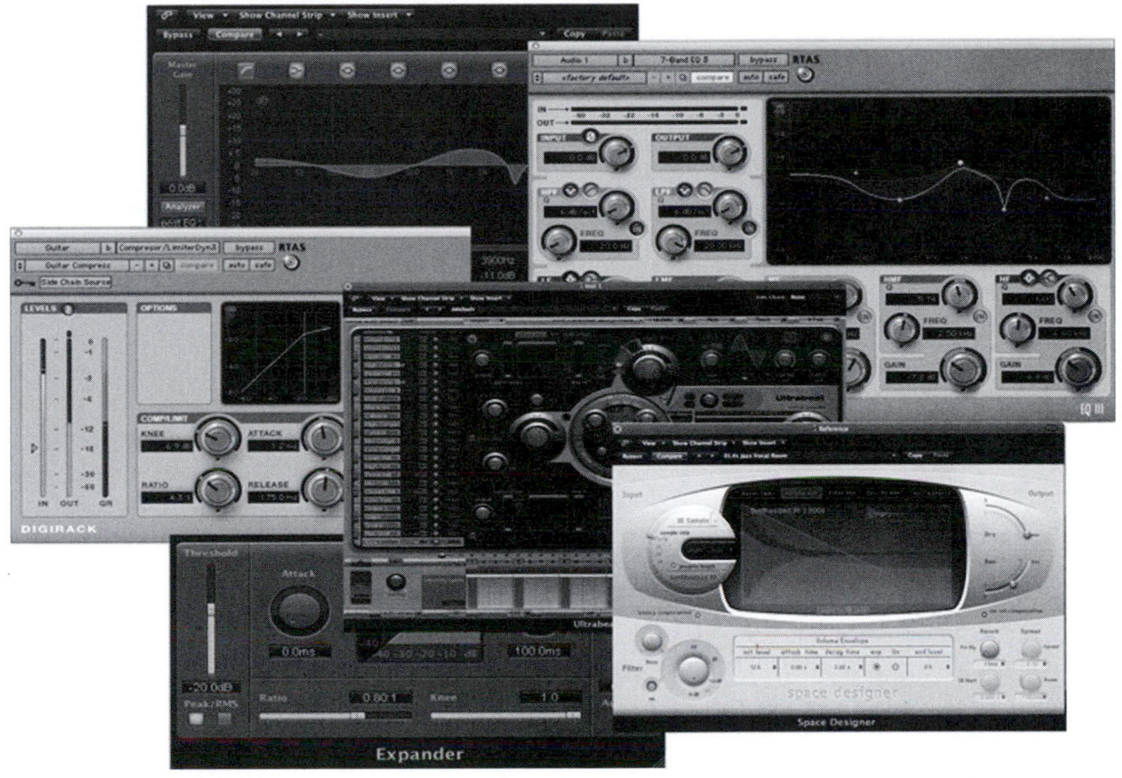

Figure 4.17
Examples of stock plug-ins from Digidesign Pro Tools and Apple Logic Pro. These plug-ins come with their respective programs and cannot be purchased separately or be used with other DAW software programs.

a way of expanding a composer's sound palette and might become quite large and extensive over time. Some virtual instrument manufacturers include Arturia, East West, and Native Instruments. The majority of virtual instruments are in the $300–400 range but can climb into the thousands for large high-quality libraries.

Third-party plug-ins are made to be compatible with each program's proprietary plug-in format. Some programs host their own format; others utilize an existing format. The chart shows current formats and the programs that are compatible. Almost all plug-ins are compatible with the formats listed, but a few are only compatible with one (because they are made by the company that makes the DAW), and others simply don't support all the formats.

A large number of plug-ins have been developed for the VST format. This is because for a long time it has been an open format with free

Plug-in Formats Chart

Plug-in Format	Compatible Program(s)
Audio Units (AU)	Logic Pro, Digital Performer
Real Time Audio Suite (RTAS)	Pro Tools (LE/HD)
Virtual Studio Technology (VST)	Cubase, Sonar, FL Studio, Live
Direct-X	Sonar
MOTU Audio System (MAS)	Digital Performer

Figure 4.18
(Created by Ben Harris.)

development software, so anyone can create VST plug-ins. So, anyone with a computer, even with no knowledge of digital signal processing or audio, can make a VST plug-in and post it for free on the Internet. For that reason, there are a lot of free VST plug-ins floating around. Some of the free VST plug-ins are totally amazing, but many of them are completely horrible. But it is often worth weeding through the many lackluster ones to find those free gems.

Modeled vs. Digital

All plug-ins are created around one of two basic premises: modeled or digital. Modeled plug-ins are created to act like a hardware counterpart. The plug-in is basically designed to function and sound like a real-life hardware unit. For example, many virtual processor plug-

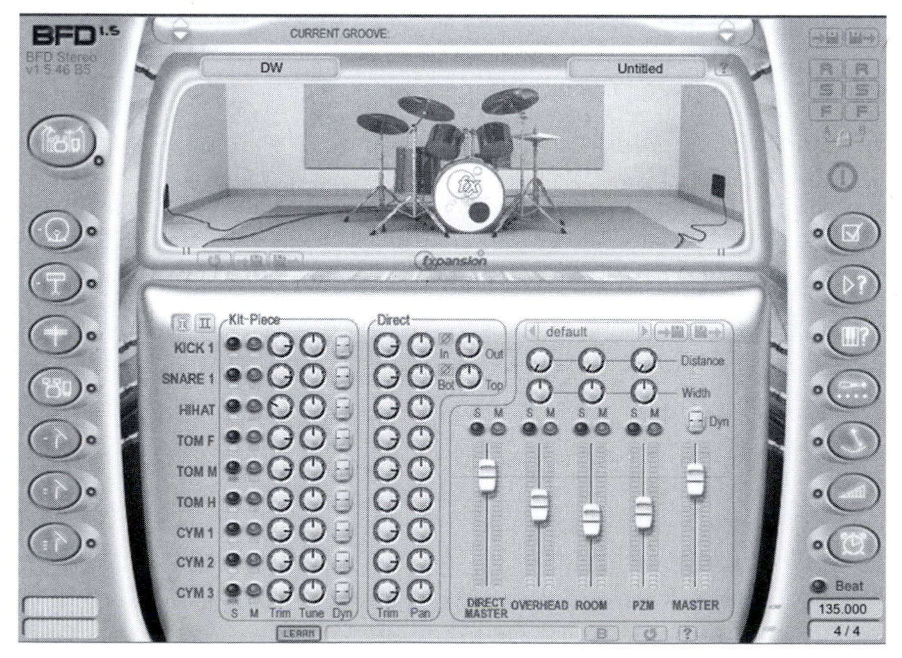

Figure 4.19
BFD, which is a drum and percussion virtual instrument. The sounds from this unit are recorded from real drums and percussion instruments, additionally providing control over microphone placement, individual instrument levels, and variations in performance dynamics and patterns.

ins are designed to sound like vintage analog compressors and equalizers. In these circumstances the manufacturers run tests and make measurements of the analog units. They then program the plug-in, test and listen, reprogram, tweak, and so on until they feel that the plug-in version sounds, looks, and acts like the original. Virtual instruments modeled from synthesizers, acoustic pianos, and drums are created through a similar process. Digital plug-ins are virtual processors and instruments that are not necessarily modeled after a specific hardware device but are created to function without the limitations of the hardware world. This often gives developers the opportunity to create a perfect equalizer, a look-ahead limiter, or a synthesizer that could not have functioned in the analog world.

Native vs. DSP

The last delineation found in plug-ins is based on where they receive their processing power. First, the majority of plug-ins are native or host based, meaning that they utilize the internal, or native, processor of the computer for power. This is how the stock plug-ins and most additional third-party plug-ins function in almost all software DAW programs. Second, there are digital signal processing (DSP) based plug-ins, meaning that dedicated processing chips, usually found on a PCI card or expansion box, provide the processing power for the plug-ins. Pro Tools HD systems are completely DSP based. Not only are the plug-ins powered by dedicated DSP chips, so is every other function, from playback to routing audio. A handful of companies provide hardware DSP cards and plug-ins that run off their power. These systems can be purchased as third-party plug-ins for any of the software DAWs.

MIDI KEYBOARDS, SYNTHESIZERS, AND SAMPLERS

In the world of virtual instruments there three main types: samplers, synthesizers, and loop players. However, the heart and soul of any virtual instrument is the MIDI that is driving it.

The most common way to create MIDI information is with a MIDI keyboard controller. MIDI is covered in greater detail later in the book, but for a moment let's take a preliminary look at how it works.

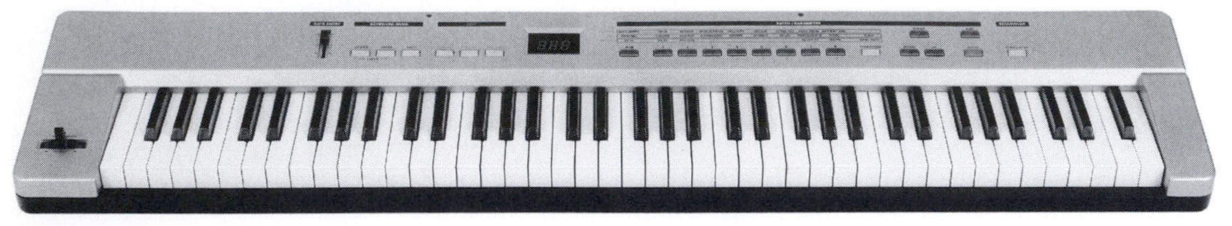

Figure 4.20
This MIDI controller keyboard includes touch-sensitive pads and is used to control virtual instruments in a software computer program. (Photo courtesy of iStockphoto, Oleg Prikhodko, Image #1591800.)

MIDI stands for *Musical Instrument Digital Interface,* but it doesn't really mean anything to simply regurgitate that useless piece of information. MIDI is a protocol. It is a system of commands (not sounds) that say that a note has been played. More specifically, there are four commands: note on, stating the time that a note begins; velocity, saying how hard the note was hit or struck; pitch, defining what note is being played; and note off, stating that the current note has stopped being played. It is then up to the virtual instruments to receive those commands and assign them to a specific sound, such as a saxophone or a piano.

MIDI Controllers

A MIDI controller is a device that creates and sends MIDI commands. Controllers can come as drum sets (triggers for acoustic drums or digital drum sets), pads, guitars, or flutes (blown and woodwind controllers); but most commonly they are keyboards. A MIDI controller keyboard does not necessarily produce any sound; it simply sends commands about what note is played, how hard, what time, and how long, as mentioned earlier. These keyboards can be as basic as a row of keys to very complex, with additional sliders, knobs, and pads. A full keyboard has 88 keys (just over seven octaves), but the majority of controllers are smaller, having 61 keys (five octaves), 49 keys (four octaves), and 25 keys (two octaves). Any of these options work fine, depending on how many octaves are needed and how much space is available in the studio.

Figure 4.21
A MIDI drum set that can be used in the studio or in live performance. (Photo courtesy of iStockphoto, Peter McKinnon, Image #7009480.)

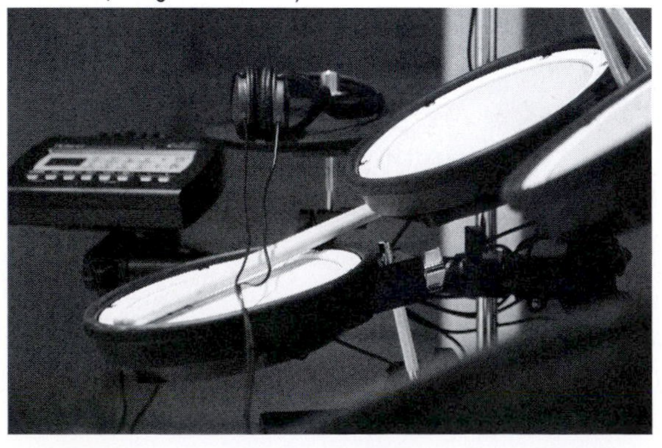

Controllers usually come with two output connection options: MIDI Out and USB. If you already have a MIDI In on an audio interface, you can use that, but the USB connection can be very convenient. The USB connection allows a keyboard to directly interface with the computer. When set up properly, the computer (and the MIDI sequencing program) will see the incoming MIDI information from the keyboard. At that point, the MIDI data can be recorded or routed directly to a virtual instrument, where it is assigned to a sound to then be heard as audio. If you play keyboards and you want to utilize MIDI, you will want a controller keyboard. It is possible to enter each note into a MIDI sequencer using a mouse, but this can be very meticulous and time consuming.

Samplers

A *sampler* is a virtual instrument that plays sounds that were recorded or sampled from real instruments. For example, if I recorded every note on a guitar and loaded those audio files into a sampler, I could play a note on a keyboard and have that sampler play back the recorded guitar note. By recording many samples of various notes, stylings, and levels of intensity, you can legitimately play a real-sounding instrument with a MIDI controller.

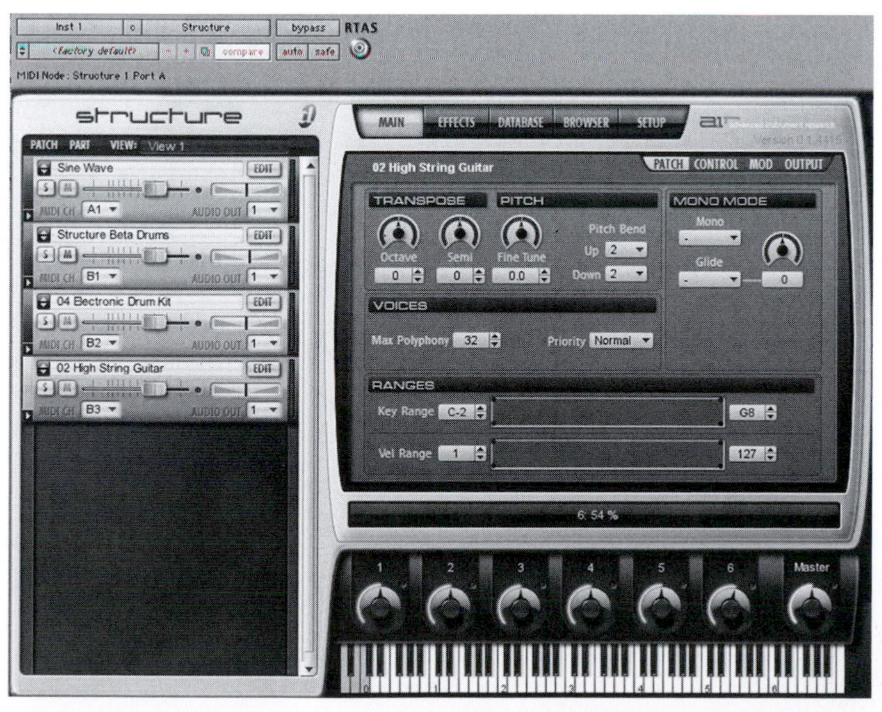

Figure 4.22
Virtual samplers like this one let you load samples of instruments or create your own.

All samplers allow you to load patches of samples to play back, but a handful let you create and load your own samples from any recorded audio that you want. These devices can be amazing tools in music production. (For more information on sampling, check out *Audio Sampling*, by Sam McGuire.) Samplers originally functioned as hardware units, allowing you to load one patch of samples (such as a violin sound) at a time. The limitation was that to have both a violin and a saxophone playing simultaneously, you needed two samplers, and each cost about $3000. Now with a powerful computer you can have multiple instances of your virtual sampler and thousands of high-quality sounds available for a fraction of the cost.

An important issue to consider when you're using a sampler is RAM usage and the alternate idea of "direct from disk" or "plug-in streaming." Most samplers traditionally load all the samples for a patch or sound into the computer's RAM when it is selected. At this point every time that you play a note, the RAM is playing that file back. The problem with this is that your RAM gets filled up very quickly. If you have too many patches pulled up, and on multiple devices, your RAM will run out of steam.

The usual solution is to simply put a ton of RAM in your computer. Even with the amazing power of current computers, many operating systems and hardware configurations can support only a limited amount of RAM. One company solved this problem by avoiding RAM almost altogether. This feature was originally called *direct from disk* and is now often referred to as *plug-in streaming* or other names, depending on the manufacturer. This feature keeps the samples on a hard drive and pulls them directly from the disk rather than load them in RAM and pull them from there. This technique usually requires a fast and dedicated hard drive just for playing samples. The benefit of this feature is that you can play back more patches simultaneously because they are not eating up your RAM.

Synthesizers

A synthesizer builds and carves sound from a generic or complex waveform by adding and taking away frequencies; modulating the signal; and changing its attack, decay, sustain, and release times. Hardware synthesizers do this with a series of analog modules that wear out, fall out of calibration, and sometimes simply break. Virtual synthesizers are either modeled after the hardware units or created

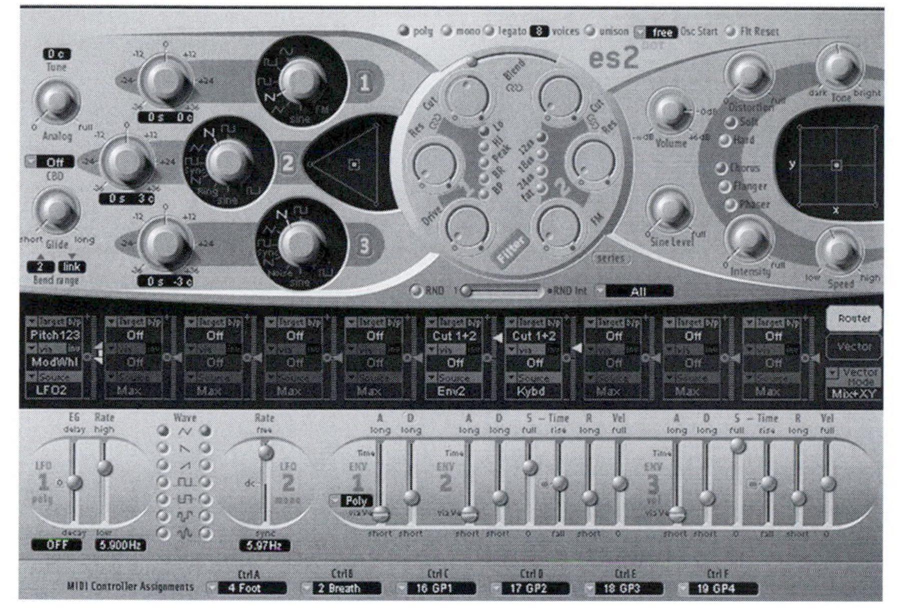

Figure 4.23
Virtual synthesizers like this one come in all shapes and sizes.

digitally. These devices let you have multiple source waves, add multiple filters, and tweak to your heart's content. The best part about virtual synthesizers is that they come with presets, so anybody can pull up a preset, maybe tweak some knobs, and get a cool synthesized sound without knowing anything about synthesizers. With hardware synthesizers you have to know quite a bit to even get them to work (let alone make a cool sound). Virtual synthesizers represent the best of both worlds; they are easy to use for the novice and customizable and tweakable for the experienced synthesist. (For more information on synthesis, check out *Analog Synthesizers*, by Mark Jenkins.)

Loop Players

Loop players are somewhat similar to samplers (especially the older hardware samplers used to create early hip-hop and rap music). The main purpose of a loop player is to play back a loop in tempo with the current project. A *loop* is a recorded instrument or sound that is usually exactly one, two, or more measures so that it can be copied and pasted after itself to form a cohesive and consistent playing performance. Loops come in the form of almost any instrument imaginable but are most common and useful as drums or percussion.

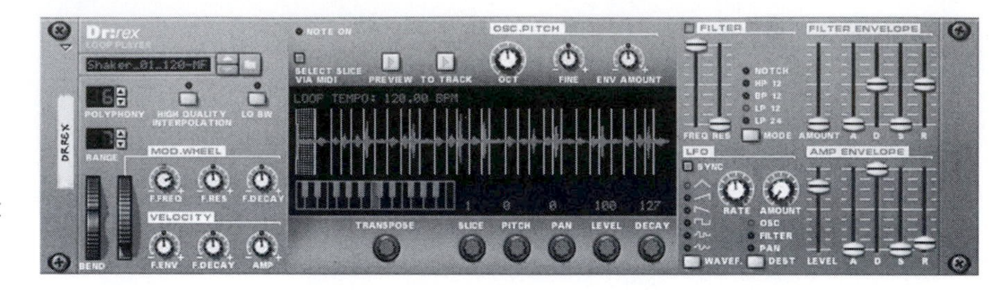

Figure 4.24
Loop players like this one let you load, play back, and manipulate loops.

Most loop players give you power to adjust the loop's tempo, pitch, and order of events. They also let you go back and forth between loops, add effects, and maybe perform scratching sounds. Loops are becoming a huge part of today's style of music production. Even if they are not heard on the final mix, loops are often placeholders until a real musician is recorded playing the part. Loop players are usually found as virtual instrument plug-ins, but a lot of DAWs incorporate the functions of loop players inside the software to facilitate the music-making process.

CHAPTER 5

The Front End

Front end is a term that has grown to a level of great importance in the project and professional studio recently. The term is based on the idea that once a source (instrument, voice, or synthesizer) is recorded, it will stay in the digital realm for the rest of its existence. Therefore it is important to get the best sound quality possible and benefit from any desired analog processing on the front side of the recording process. There are multiple items that can be part of the front end. The first element is the microphone, followed by a microphone preamplifier, optional extra analog processing such as equalization and compression, and analog-to-digital conversion; finally everything is connected together by cables.

I would rank these in importance based on the idea that a more costly and therefore higher-quality piece of equipment will give you better sound quality, in the following order:

1. Microphone preamplifier
2. AD (analog-to-digital) converter
3. Microphone
4. Cabling
5. Extra analog processing

The equipment of the front end is constantly changing with better pricing conditions and new technology. Keep current with prices, reviews, and audio samples on theDAWstudio.com, the companion website to this book.

TRADITION

Traditionally the front end was well taken care of by a high-quality recording console and an extensive microphone cabinet. Each channel of a high-end console had a high-quality microphone pre-

Figure 5.1
A modern analog console that provides a high-quality front end for tracking. (Photo courtesy of Immersive Studios, Ben Harris.)

amplifier, an equalizer, and possibly dynamics controls (compression, gating, de-essing). This is why a 64-input console costs a few hundred thousand dollars.

As studios got smaller, they wanted the quality of larger studios without the huge expense and footprint of a large console. People began to pull channel strips (microphone preamp and equalizer) out of old consoles and put them in racks that provided power and analog connections. This meant that you could have the sound quality of a professional studio at home. Eventually companies began building standalone microphone preamps and channel strips in the $2000-plus price range. Now there are middle-quality products starting as low as $100 as well as high-quality ones moving up into the thousands of dollars. The general idea is that by having good processing on the front end, you will get the best signal possible to work with later on, during editing and mixing.

MICROPHONES

First in line is the microphone, the element that captures the sound of an acoustic instrument, voice, or amplifier. There are three main types of microphones used in recording: dynamic, ribbon, and condenser mics. These types each capture sound in a different manner, so they each have a different sonic character, responsiveness, and ideal application. Microphone quality can vary greatly, with prices ranging from $50 to $5000. With advances in manufacturing, technology, and a ribbon renaissance, good-quality microphones have become more and more affordable to the project studio.

Figure 5.2
A Shure SM57 dynamic microphone, commonly used for recording snare drum and electric guitar. (Photo courtesy of iStockphoto, Dave Long, Image #4431924.)

Dynamic Mics

A dynamic microphone can also be referred to as a *moving coil*. The diaphragm is connected to a metal coil that is suspended in the center of a magnet. As the diaphragm moves back and forth, the coil moves in and out of the magnetic field, changing the polarity of the coil and creating an alternating current on the wire that is connected to the coil. These microphones are very robust because of their simple design. They are used quite extensively in live performance as well as for drums and electric guitars in the recording studio. One of the most common dynamic microphones is the Shure SM57. This microphone is used on 90% of the snare drums heard in popular music and at least 50% of distorted electric guitar. It is also seen with its sibling, the SM58 (a vocal mic), in almost every live performance.

The sonic character of a dynamic microphone is usually defined as being more prominent in the upper midrange, with a somewhat grainy and less prominent high end. These have an interesting responsive character that makes them useful for drums, guitars, and basically more explosive content. Dynamic microphones take some inertia to get going because the diaphragm has more mass and weight to it. This extra effort or energy needed to get the microphone going helps recreate explosive material because you can almost feel the movement of energy or the effort that the sound has to go through to be heard. Recently dynamic microphones have progressed in leaps and bounds in terms of technology. There are now many dynamic microphones that have the clarity and smoothness of a condenser mic.

Ribbon Mics

A ribbon mic can actually be referred to as a dynamic microphone because it captures sound in a similar way, without the use of external power. Instead of a coil connected to a diaphragm, however, the element being acted on is a very thin strip of folded aluminum (or other thin metal). This ribbon strip is placed between two magnets in a field that is disrupted as the ribbon moves back and forth and dances around in reaction to the sound waves in the air. Once again, this creates an alternating current, which is sent down the wire.

Figure 5.3
A vintage ribbon microphone, commonly related to recording during the 1940s and 1950s. These microphones have recently made a comeback. (Photo courtesy of iStockphoto, David Lentz, Image #2394240.)

Ribbon microphones are much more delicate than moving-coil microphones because of that thin aluminum ribbon. These were originally used quite extensively in broadcast during the first half of the 20th century. Two of the most famous ribbon microphones are the RCA 44 and 77 (shown in Figure 5.3 on the previous page), made by a broadcast company. There has recently been a "ribbon renaissance," with new ribbon designs by countless companies, bringing back a classic sound and sometimes creating a modern ribbon sound.

The sonic character of a ribbon microphone is usually referred to as having a rolled-off high end, but that doesn't seem very attractive when you say it that way. In comparison to the character of a condenser microphone, ribbons do not have the same hyped high end. They have a much smoother and natural-sounding high end, more like what we hear in real life. Ribbon microphones also react very differently based on how far they are placed from a sound source. The high end does seem more rolled off when used in close micing, but when pulled back there is a smooth presence of crisp and clear high frequencies. Ribbons are often used in distant micing, such as drum overheads or room mics, guitars, horn instruments (because they smooth the sometimes brash highs), and vocals.

Condenser Mics

A condenser microphone is based on a moving plate and a charged fixed plate and the capacitance that is generated between them. The diaphragm of a condenser microphone is very thin and sputtered with metal (usually gold) to make it act as a metallic plate. There is a back plate placed behind the diaphragm with a small distance between them. Both plates are charged with an electrical current, and as the moving plate (diaphragm) moves, capacitance is generated between the plates and an alternating current is created, reflecting the changes in capacitance. The output of this process is at a very low level, so condenser microphones require power for both an internal preamp to boost the low level signal and to charge the plates. This can be provided by an external power supply, phantom power, or a battery. External power supplies are still used for tube microphones, which have a tube inside the micro-

Figure 5.4
A condenser microphone, commonly used for vocals and just about any instrument in the recording studio. (Photo courtesy of iStockphoto, Mike Bentley, Image #445143.)

phone and use a special cable to transmit both audio and power to and from a power supply unit.

Originally all condenser microphones had power supplies until a microphone manufacturer (by the name of Neumann) discovered that he could send power to the microphone on the same cable that sent the audio signal from the microphone. Phantom power sends up to 48 volts of electricity up the wires on an XLR cable from a microphone preamp to the microphone. This is supplied by the microphone preamp of a standalone unit or from a console. Phantom power does nothing to dynamic microphones but can severely damage ribbon microphones. There are now some battery-powered microphones to use in low-power situations.

The sonic character of a condenser microphone is defined as having a hyped or prominent high-frequency response, with a bump (in large diaphragm condensers) related to the resonant frequency of the diaphragm. This bump or high-frequency presence boost is often what gives a certain condenser microphone its character. Small diaphragm condensers still have this bump, but it is generally not noticeable, because it is at a much higher frequency (in relation to a smaller diaphragm size and therefore higher resonant frequency). Condenser microphones are most commonly used in the recording studio, although they are found more and more often in live settings as well. Condensers are characterized by their amazing detail and clarity (because of the thin diaphragm). They can be used on anything in the studio, including drum overheads, acoustic guitars, and vocals.

The latest rave in manufacturing has been to use Chinese-made condenser diaphragms in microphones. This does produce quality microphones at a cheaper price, but there is often a harsher high end. The expensive European-made diaphragms and microphones still have a hyped but very smooth and silky high end.

Microphone Polar Patterns

Polar patterns show the directionality of a microphone's pickup pattern. The graphs shown here are an aerial view of a microphone, with 0° representing the direction in which the front of the microphone is facing. All three patterns pick up a strong portion of their signal directly to the front. It is how much signal they pick up from the back and sides that poses the issue. There are three main patterns

POLAR PATTERNS

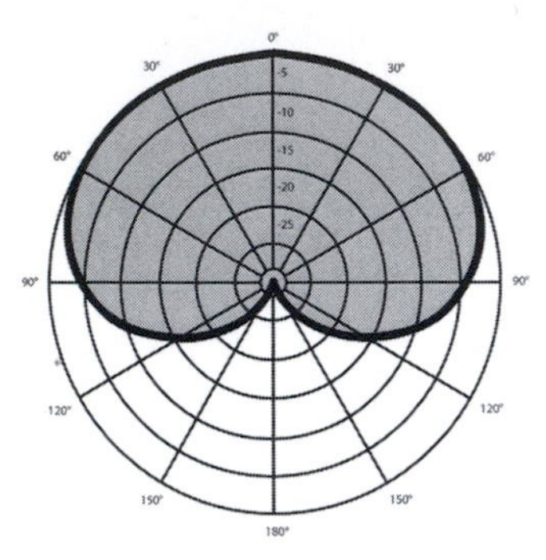

Cardioid
(Unidirectional)

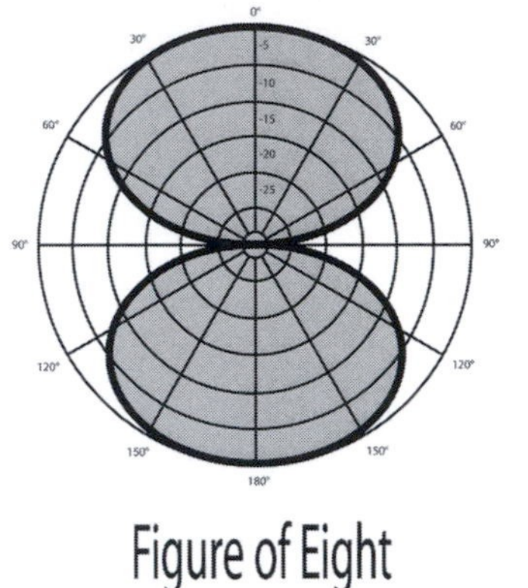

Figure of Eight
(Bidirectional)

Omni
(Omnidirectional)

Figure 5.5
The three main types of microphone polar patterns, with 0° representing the front of a microphone. (Created by Ben Harris.)

as well as variable patterns in between. The three main patterns are cardioid, figure of eight, and omni.

The first pattern, *cardioid,* is the most common. This pattern picks up information in front of the microphone and rejects information from the back and sides. This pattern looks like an upside-down heart (thus the name cardioid). Almost all dynamic and condenser microphones have cardioid patterns. This pattern is created by permitting information to reach the back of the diaphragm out of phase, canceling itself out. This directionality helps avoid picking up unwanted sound, focus on specific sources, and avoid feedback.

The second pattern, *figure of eight* (or *bidirectional*), is less common than cardioid but still very useful. Figure of eight is the inherent pattern of most ribbon microphones and key to the m/s stereo technique. Figure of eight picks up information equally from the front and back of the microphone. Each side picks up the opposite phase of the other (not out of phase) because air is pushing it from opposite directions. This pattern is very useful for duets when singers are placed on either side of the mic. It is also good for close micing, picking up a close sound in the front of the microphone and more room acoustics from the back.

The third pattern, *omnidirectional* or *omni,* is mainly available on condenser microphones. It picks up information equally from all directions. This pattern is the best sounding but most difficult to work with. Omni sounds the best because it picks up an even balance of frequencies from every direction. The other patterns do not look the same at all frequencies (for example, the cardioid pattern gets more directional with higher frequencies and more omnidirectional with lower frequencies, creating a varying frequency response from different angles). The problem with omnidirectional microphones is that they pick up everything, so you have to have a good-sounding acoustic space to really use an omni. They are great for close micing (because they don't have a proximity effect, which we'll discuss later) and they're excellent for room microphones.

There are also a few patterns in between these three. *Hypercardioid* is narrower than regular cardioid, and *supercardioid* is in between *hypercardioid* and *cardioid.* Some multipattern microphones are completely variable, slowly transitioning through the patterns and everything in between. Polar patterns are an important aspect with microphone choice and placement considerations (discussed in Chapter 9).

MICROPHONE PREAMPLIFIERS

The microphone preamplifier is next in the chain of an audio signal. It is the most important step because the signal coming from a microphone is at a very low level and needs to be amplified to a useable level (line level). If the amplification of the signal is done poorly, the signal will sound grainy, with poor dynamics and rolled-off high and low frequencies. If the amplification is done well, it will make any microphone sound amazing.

Figure 5.6
A rack full of high-quality external microphone preamplifiers. (Photo courtesy of Immersive Studios, Ben Harris.)

Basically, the idea is that you will get a much better-sounding signal if you run a $100 microphone through a $2000 preamp than if you run a $2000 microphone through a $100 preamp. Of course, a $2000 microphone running into a $2000 preamp would sound the best of all three scenarios. This is why microphone preamps are such a hot topic with recording engineers. Home studio users go on about which microphone is

DISCRETE

VS

IC (Integrated Circuit)

Figure 5.7
This image shows two audio processors, one with discrete electronics and the other with integrated circuits (ICs). (Photos courtesy of Wind Over the Earth, Ben Harris.)

more expensive and cooler than the rest, but the real thing that they should worry about is the preamp.

There are some basic features that differentiate the consumer and prosumer (products that are in between professional and consumer quality) level of preamps from the professional ones. First there are integrated circuits (ICs), which basically have multiple electronic circuits built into a single chip. IC's are the reason that we can have so many great electronic products in small packages, but when it comes to high-end audio they just don't sound as good as discrete electronics. *Discrete electronics* basically means that every electronic circuit is individually found on the circuit board (that is, if the signal needs to go through a capacitor, there is a capacitor on the board in the signal path). This is how all analog equipment was designed before ICs, which is why older preamps and microphones sound so good. Now we have the option to make our equipment small and cheap, which is both good and bad.

Second are Class A electronics. This has to do with how the preamp amplifies the signal. Class A amplification is less efficient (it uses more electrical power to do less work), but the output is a near-perfect representation of the unamplified input, with hardly any distortion. There are also Class B and A/B amplifiers, which are all more efficient but have higher amounts of distortion in the process.

Consequently there is a price difference between consumer- and prosumer-level preamps and professional preamps. Professional-level preamps start at $500 per channel at the cheapest. This dollar amount works fine for single-, dual-, and quad-channel preamps, but there are some professional level eight-channel preamps in the $3000 range (which throws the number off a bit because of quantity pricing). Anything under this price range may sound decent but will not have the clarity, dynamic range, and frequency response of the professional-level units.

Tubes vs. Transformers (Solid State)

There is also the issue of tube versus solid-state electronics in microphone preamps. This debate is not similar at all to that involving guitar amplifiers. Many people have the idea that a tube microphone preamp will have a "fatter" sound. This is simply not true. The "fat" sound in microphone preamps usually comes from the transformers. Even in tube equipment, the transformers surrounding the tubes often define the sound. There are many great tube preamps that sound colored and many that sound ultra-clean. The same thing goes for transformer-based (solid-state) preamps.

Figure 5.8
(Photos courtesy of iStockphoto, Jakub Semeniuk, Image #5164615; Daniel Brunner, Image #398242.)

There is a division in high-quality microphone preamps between clean and colored. As mentioned earlier, tube and transformer-based preamps can be on either side of this line. The clean preamps are based on the idea of a wire with gain. The goal is no coloration, with the closest representation to what is coming out of the microphone as possible. Some of these preamps avoid both tubes and transformers to evade any coloration of the signal (usually called transformerless). This is very useful in choral, instrumental, and orchestral music. These preamps give you pristine clarity, amazing depth and imagery, and a flat extended frequency response. The colored preamps give character to the signal, usually based on the sound of the audio running through various transformers or tubes. These are often used in popular music to make drums sound larger than life, push guitars right in your face, or make a lead vocal wrap itself around you. The sound can have thousands of different characteristics, including frequency boosts or dips in various

Figure 5.9
Outboards equalizers and compressors that could be used on the front end or during mixing. (Photo courtesy of Immersive Studios, Ben Harris.)

parts of the spectrum, increased dynamic range, and fast or slow transient response. One key difference between the colored and the clean is that the colored preamps are clear, creamy, and detailed but not necessarily pristine.

OUTBOARD EQUALIZERS AND COMPRESSORS

Analog processing on the front end is, as you might have noticed in the beginning of the chapter, listed at the bottom of the priority list. This is because the current tendency, and my preferred method, is to bypass any additional processors and go straight from the mic-pre out to the AD converter input. There are a few good reasons for putting some analog processing, such as equalization and compression, on the front end. One is to make your signal sound more like the original source; the next is to make some mixing choices during recording; and finally it is done to put some rich and pure analog goodness on the signal before going into the digital domain.

Although it can be for any of these three reasons, it can also be any combination of the three. There are many typical processors that can be used in this manner. These include equalization, compression, de-esser, and gate. These processors can be found on the mic-pre, channel strip, or individual units. The first purpose for adding processing on the front end is to make the signal sound more like the original. This reason can be somewhat controversial because it is best to utilize microphone choice, room positioning, microphone placement, and microphone preamplifier choice to accomplish this task. Ideally you can make your signal sound practically identical to the way the original sounds in the room if you have an infinite (or quite expansive) list of choices of microphones and preamplifiers. I imagine that not everyone reading this book has that many choices, so adding processing could get the signal sounding a little closer to the original.

The most commonly used processor in this circumstance is equalization. High-pass filters are often found on microphones and preamps to get rid of air-conditioning rumble, mechanical noise, and unwanted proximity effect. Two- or three-band equalizers are standard on console channel strips and can be used to make the signal a little brighter, get rid of middle muck, or boost the lows. Compressors can also be used to make the signal sound more like the original by limiting some of the transient response of the microphone, taming some of the more prominent frequencies by compressing them first, or simply by rounding out the dynamics of the sound.

The second purpose for adding processing on the front end is to make mixing decisions during recording. These kinds of changes are usually best done by someone experienced in mixing who knows what individual elements need to sound like in a completed mix. This can include making some changes with equalization, such as cutting some of the lows from an acoustic guitar, because they know that if they don't do it now they'll just have to do it later. Compression or limiting can be added to vocals or bass when you know it will be necessary when you're mixing. De-essing can be used to fix a sibilant vocal before it gets to the mixing stage. Gating is commonly used on drums with the close mics to limit the amount of bleed recorded while the individual drum is not being played.

Other processing such as time-based effects can be done at this time as well. Adding processing for this purpose is often scary, but it helps in not putting off decisions and letting music just … happen.

If you want to take a more conservative and modern approach, you can actually split the output of a microphone preamp with a patchbay or specialty cable and send one signal to be recorded and the other to be processed and then recorded. Then you have two tracks recorded, one with processing and one without. That way if the processing turns out to be a poor choice or excessive, the original can be used instead.

The third purpose for adding processing on the front end is to put some analog goodness on your signal before it gets digitized. This is something that a lot of professionals do with their old vintage equipment. Something beautiful happens when a signal goes through certain analog circuits. The signal is slightly changed when it goes through different transformers, tubes, or circuits. The color that is added is often what makes the sound of a track. There are even boxes that simply send a signal through heavy transformers, no EQ or compression, just to color the sound. This is similar to the use of different preamps to get different colors.

One example is the Urei 1176. For years this compressor has been used on hit tracks. Most of the time it shines for its smooth compression and limiting, but it is often set to bypass so that the transformers color the sound but the signal is not compressed. Another example is analog tape compression. When a loud signal goes to analog tape, it begins to distort. A slight amount of this type of distortion can give drums extra color and punch. Engineers will often record their drums onto tape and then to a DAW just to get this sound.

Using processors on the front end, such as equalization, gating, and compression, can be looked at as nonessential or a luxury. It is something that beginners shouldn't overdo until they have more experience with those same processors in the mixing environment. Many people suggest using a limiter before your signal gets to the converter, to prevent digital clipping. If set up correctly, this can be a wonderful safety net in case a performer suddenly gets louder than expected during a performance. Many converters have built-in limiters just for this feature. I suggest using them as a safety net if they're there, but the danger in suggesting that everyone put a limiter in their signal chain is that people are bound to misuse it and compress their signal while recording, then regret it later.

ANALOG-TO-DIGITAL CONVERSION

After the signal has been captured, amplified, and processed in the analog domain, it needs to be converted to digital information to be recorded in the digital domain. The details of analog audio in comparison to digital audio are covered in Chapter 8. The key thing to discuss now is the quality of the conversion at this point in the front end. Just like microphone preamplifiers, AD converters vary greatly in quality. There are a few key concerns you should keep in mind when choosing what converter to buy, including price, clocking, and jitter.

Figure 5.10
An analog-to-digital (AD) converter. A digital-to-analog (DA) converter looks very similar, only with analog outputs instead of inputs. The front panel contains a handful of options to control bit depth, sample rate, and clock source. (Photos courtesy of Wind Over the Earth, Ben Harris.)

There are converters on the built-in sound card of a computer, and there are converters that cost $7000 for two channels. There are audio interfaces that incorporate mixer functions and conversion that also range across the same gamut of price. Unlike many other elements in audio, converters have a pretty equal ratio of price to quality. Basically, you get what you pay for. My advice on the low side is, first, do not purchase an interface/converter to use for recording from a store that is not a music or recording equipment retailer (that is, don't buy one from a computer store). Second, don't pay any less than $250 for two channels of conversion. On the high side, the price points set forth earlier for microphone preamplifiers ($500 per channel) are basically the same minimum for the starting point of high-quality converters. The clock is really the most important aspect of a converter. Conversion is based on so many samples per second, so the clock, which dictates how fast and consistent the seconds are going by, can have a great affect on the quality of the outcome. Each converter has a built-in internal clock. The quality of the clock is a huge part of the converter quality, so good converters have good clocks and excellent converters have excellent clocks. A solid clock will give the signal less jitter, more clarity, and a deeper image. A good clock can also be a significant upgrade for a mediocre system. If you tell a good converter to ignore its clock and look to an excellent clock, it will sound like a "better than good, but not quite excellent" converter.

There are many things that a converter has to do and hopefully do well, such as up-sampling, filtering alias frequencies, and dithering. All these tasks can have an effect on the sound quality of the converter. One of the most noticeable side effects of less than excellent quality conversion is jitter. Jitter is related to clock stability, but it is also related to some of the previously mentioned tasks of a converter. Jitter is noticeable as washiness in the sound or clicks in higher frequencies. Jitter occurs when the clock speeds up or slows down and samples are either doubled or skipped to keep in time with the clock. Jitter is something that you will hear about in reviews or technical papers on specific converters.

The AD converter is a key component in the quality of the front end. Just remember to consider that the clock is key to a good-sounding converter; jitter is what you should avoid and will be minimized as you get higher quality conversion and that you get what you pay for when it comes to converter price.

CABLES

Some people will tell you that the cabling used to connect all the elements of the front end is the most important aspect. They are both right and wrong. The key thing to remember is that your signal quality can only be as good as your weakest link. So, if you have a $5000 microphone going into a $2000 microphone preamp going into a $3000 AD converter and they are connected with $5 cables, your signal will only sound as good as the $5 cables can make it. You will basically lose clarity and high-frequency response. The elements that need to be considered with cables are the material the contact points on the connectors are made of, the material the cable is made of, and how well the cable is shielded.

Cable Materials

There are a few things to consider with a cable connector. First, is it easy to plug in and unplug? Is it ergonomically comfortable to make connections? Second, if you are going to build your own cables, is it easy to solder? Is it simple to put together or take apart? Third, and most important, how much contact area is there, and what material is it made of? When it comes to conductivity of an electrical current, the most common metal used is copper. The problem with copper on a connector is that it corrodes. Most cost-effective connectors are nickel or nickel plated, which is noncorrosive but not a great conductor. Gold or gold-plated connectors are the best option because gold is noncorrosive and a good conductor. The problem is that, as we all know, gold is expensive, so these higher-quality and better-conducting connectors are going to cost more.

Cable material and quality are very important issues. There are two types of cores to cables: solid and stranded. A solid core conducts electricity better because of surface area, but it is stiff, difficult to work with, and can break if it is moved too much. These cables are excellent for fixed installations. The second type is stranded or braided. This core is made up of a braid or multiple strands of smaller cable. This is what audio cable commonly consists of because the cabling needs to be flexible and durable.

The next thing to consider with the quality of a cable is the material the core is made of. Just as with connectors, some metals are better than others for conducting electricity. Copper is the most common conductor because it is one of the best conductors of electricity and is readily available and fairly inexpensive. The best conductor of electricity is silver, which is used in some high-end audio cables, but it is very expensive and is often not worth the price difference. Silver is not necessarily the way to go for high-quality audio. The majority of professional recordings are made with copper cabling. Both silver and copper are corrosive, but they are covered and therefore not exposed to moisture and the air.

Shielding is the third key component to quality cabling. If a cable is not well shielded it can experience interference from radio frequencies, power cables, and other audio cables. There are a couple of techniques for shielding. One is to have the ground wire braided around the insulated core wire. This is very commonly used and works great. Another technique is to place a foil shield over both the ground and core wires. The aluminum shield is more efficient and is often used on snakes if you have multiple small cables bunched together. When a cable is used a lot, the aluminum shield can wear out and crack, making it useless. A braided shield can also wear out from overuse, but it takes a lot more time to wear out than the aluminum shield. There are other techniques that companies use to thoroughly shield and insulate cables, but the main idea is to make sure that the cable you buy is properly shielded.

Overall, cables are very important. I wouldn't suggest going crazy and using silver cable on your entire studio, but I would suggest budgeting a few hundred dollars just on cables.

Cable Types

There are a few main types of cables and many variations in terms of connectors, number of contact points, and impedances. On a basic level there are six major cable formats: line, microphone, instrument, speaker, digital, and video. These six formats have common connectors between them, so it gets confusing at times. These formats differ in signal level, connector type, and electrical impedance. The impedance has to do with the interference between individual wires of a cable or the multiple signals from a device.

First, line-level signals are sent over many different connectors and at two different levels. These two levels are balanced at +4 dBu and unbalanced

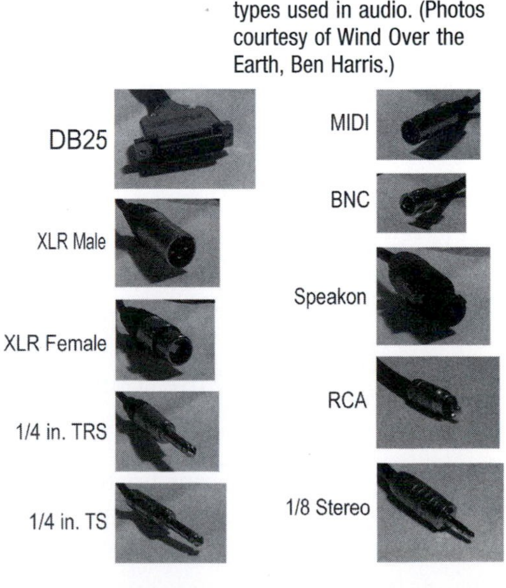

Figure 5.11
Many of the cable connector types used in audio. (Photos courtesy of Wind Over the Earth, Ben Harris.)

at −10 dBu (dBu is a measurement based on voltage levels of the alternating current). One difference between these two is that most professional equipment sends signals at +4 dBu (balanced) and most consumer equipment sends signals at −10 dBu (unbalanced). The terms *balanced* and *unbalanced* have to do with the way a signal is sent and then received.

Unbalanced cables have two leads (or two connections or lines) running through them. One is the signal and the other is the ground. The ground wire shields the main signal from electronic interferences that could modulate the main signal and then connects to the ground of the device that it is going to, usually eliminating the noise. Balanced cables have three connections. Two of them carry the signal, but the signal has been duplicated and put out of phase in one of the cables. One is called the hot (or positive). This is the original in-phase signal. The other is called the cold (or negative). This is the out-of-phase signal. The third signal is once again the ground (or shield), functioning the same way as in the unbalanced cables.

As the signal travels down the cable, noise is picked up equally on the two signal wires. When the signal reaches its destination, the cold (or negative) signal is put back in phase so that when added to the hot (or positive) signal, any noise picked up along the cable length on the cold signal is now out of phase with the noise picked up on the hot signal; when added together only the noise is canceled out. Balanced signals are sent at a higher level (+4 dBu) because there is less issue with noise. Unbalanced signals are sent at a lower level (−10 dBu) because they have a tendency to pick up less noise at lower signal strengths. Problems often occur when levels are mismatched. For example, if you send a balanced signal to an unbalanced input, you will often get distortion, since your signal is 14 dB hotter (or louder) than what the input is expecting. The opposite causes problems as well, giving you a low-level signal that often becomes noisy when amplified. Many units have smart inputs and outputs that will change to unbalanced or balanced based on whether the connected cable has two or three contact points.

Balanced line-level signals are usually carried over two different cable connector types, ¼-inch TRS and three-pin XLR. Both connectors have three contact points to carry the balanced (+4 dBu) signal. A ¼-inch TRS connector looks similar to a guitar cable but has an additional line on the connector, creating three contact points. A three-pin XLR connector is identical to a microphone cable (but they do not carry the same level signal) and the cables can be used interchangeably. The ¼-inch TRS and three-pin XLR carry identical signals, so the connectors can easily go back and forth or be substituted for each other, depending on the input connection. Eight channels of balanced line-level audio can also be sent over a DB25 25-pin connector (using 24 of the pins). Balanced line-level connections are used to connect external equipment (or professional keyboards) to mixers (or audio interfaces) and to connect main outputs into power amps or powered speakers.

Unbalanced line-level signals are usually carried over ¼-inch TS and RCA connectors. A ¼-inch TS cable is identical to a guitar cable (but they do not carry the same level signal) and the cables can be used interchangeably. RCA cables are the normal red and white cables used in most consumer audio equipment. Unbalanced line-level signals are commonly used to connect keyboards and unbalanced equipment to mixers (or audio interfaces), to connect a main output to a power amp or powered speaker, and to connect DVD players into receivers in consumer electronics. Stereo cables are usually ¼-inch or 1/8-inch TRS connections, but they are not balanced. They are two unbalanced signals traveling over a single cable. They have three contact points: one for left, one for right, and the ground is shared by the two signals. This is what headphones and many consumer computer sound cards use to transfer a stereo signal.

Microphone cables are identical to balanced XLR line-level cables, but the signal is very different. A microphone sends out a balanced signal, but it is not a +4 dBu signal. It is more in the range of −50 dBu to −60 dBu. This is why a microphone signal is sent straight to a microphone amplifier. The microphone amplifier brings that low-level microphone signal up to a +4 dBu line-level signal. A microphone signal also has a lower impedance that makes it susceptible to signal loss and degradation over longer cable runs.

Instrument cables are identical to line-level ¼-inch TS cables, but once again the signal is different. This time the signal is stronger than that of a microphone, but it's still not as strong as a line-level signal. The signal also has a very high impedance level. Instrument inputs are found on microphone preamplifiers, mixers (or audio interfaces), and guitar amps. These inputs are often called *direct-inject*

(DI) inputs. These inputs are designed to receive the high-impedance signal that comes from a guitar and adjust that signal to either a line or a microphone level. DI boxes receive an instrument signal and change it to a microphone-level signal so that it can be sent into a microphone preamp of a console at a live show. Since instrument cables are unbalanced, they have a tendency to get pretty noisy with long cable runs.

Speaker cables can look a lot like guitar cables, but they are very different on the inside. A speaker cable does not have the same kind of shielding mechanism as the cables we've previously mentioned. It does have a ground, which travels next to the main signal. These cables are usually thicker (or lower gauged) because they carry an amplified signal. A few different connectors are commonly used for speaker cables. The most basic is a bare wire connection in which the bare wire from the cable goes into two clips. A banana connector is very similar to this idea except it uses two parallel plugs to connect the wires. A speakon connector (shown in Figure 5.11) is hard to explain, but the key is that it locks into place.

Finally, one of the most common speaker cable connectors is a ¼-inch TS connector. These cables look very similar to guitar cables, but there are a couple of ways to tell the difference between the two. One way is to look at the size and thickness of the cable and connector. Speaker cables often have bigger ¼-inch connectors and thicker (lower-gauge) cable. The other way is to unscrew the connector (if you can) and see whether the wires are running next to each other (speaker) or if there's one cable surrounded by a shield (guitar).

Digital cables get even more confusing because they look the same as XLR and RCA cables but have different impedances. Six main digital formats are used in audio and one additional format for digital synchronization. The first is the Audio Engineering Society/European Broadcast Union (AES/EBU) format, commonly called AES. This format sends two channels in one direction over a single three-pin XLR cable. It can also be sent as eight channels in and out on a DB25 25-pin connector. You can't simply use a regular microphone cable, because the impedance mismatch will cause signal loss and provide glitches, pops, and errors.

The second format is Sony Philips Digital Interface Format (S/PDIF, pronounced *spid-if*), which is very similar to the AES format except it is sent over an RCA cable. Once again, a regular RCA cable will not work properly. Third is the ADAT optical format, which is an eight-channel signal sent over an optical cable. Fourth is the Toslink format (sometimes called S/PDIF optical), which is basically a two-channel S/PDIF signal sent over an optical cable (identical to the optical used for ADAT optical). Fifth is Tascam Digital Interface Format (TDIF, pronounced *T-dif*), which transfers eight channels over a DB25 25-pin connector. This uses the same connector as the eight-channel AES connection, but the formats and cables are completely incompatible and not interchangeable. Sixth is Multi-channel Audio Digital Interface (MADI), which can carry up to 64 channels over a fiber-optic cable. MADI is big, expensive, and mainly used with large digital consoles in big production facilities. Finally there is word clock, which is a synchronization format used to provide timing information to multiple digital devices so that they all run in synchronization. This cable is a locking coax cable that is also used in some video applications.

Video cables are a whole other world from audio, but the main cable to worry about is the RCA-type coax cable. This cable looks identical to an RCA unbalanced line and S/PDIF cable but should not be used in place of either of these. Once again there are impedance mismatches that will degrade the signal.

There are a few additional cables not mentioned here that can be used in a home studio setup. These include the MIDI five-pin DIN connector, USB and Firewire cables, and standard power and IEC (the power connection going into your computer or audio equipment) connectors.

CHAPTER 6
Control and Monitoring

Controlling the processing and mixing of your audio and monitoring what you are working on are important elements in any studio setup. There are many options as to the way you accomplish this task. Mixing consoles have been the traditional way to process audio, but now that many of those functions are handled inside the software DAW, software control surfaces have emerged as a new working environment. The mixing console has also traditionally handled the adjustments and controls of the monitoring levels of your sound, but now that many studios do not have consoles, there are many alternative options. There is also the issue of studio monitors (speakers). Why are there so many options, and why won't home theater or bookshelf speakers work? All of the equipment involved in control and monitoring is constantly changing with new technology, advanced features and more options. Keep current with prices, reviews, and audio samples on theDAWstudio.com, the companion website to this book.

MIXING CONSOLES

As mentioned earlier, traditionally a large mixing console served a substantial portion of functions in a recording studio. First, it has microphone preamplifiers to record high-quality signal from your microphones. Second, it has mixing and routing functions, including equalization and sometimes compression. Third, it has level control for speaker and headphone monitoring.

Mixing Consoles vs. Audio Interfaces

When computer recording began to emerge, audio interfaces were basically nothing but AD/DA converters ready to interface with your current mixing console. If you didn't have a large mixing console,

Figure 6.1
Both a mixing console and a software DAW are shown. Do you use one or the other, or some combination of the two? (Photo courtesy of iStockphoto, Chris Schmidt, Image #5312894.)

you had to buy a small one to function with your audio interface. Soon manufacturers began to add mixer functions such as microphone preamps, headphone outputs, and monitor-level control into their audio interfaces. Now a home studio setup can be as simple as a computer, an audio interface, speakers, and a microphone. To many people that is the ideal home studio, but professional studios have progressed as well. Many no longer have a large analog console, but they also don't have an audio interface with mixer functions built in (they require higher-quality components than those found on a prosumer audio interface). They have individual microphone preamps, audio interfaces (AD/DA converters), and monitor and headphone controllers. These component systems are very high quality, customizable, and much more compact than a large mixing console.

Analog vs. Digital Mixing Consoles

There has been a proliferation of all things digital in the world of recording, but analog is still not a thing of the past. When it comes to mixing consoles there are a pretty even number of analog and digital versions. An analog console is still similar to the consoles of the past. It has microphone and line-level inputs; equalization on every channel; inserts and auxiliary sends and returns; busses; and

Figure 6.2
(Photos courtesy of Immersive Studios, Ben Harris.)

so on. Many larger consoles have compressors that can be patched into individual channels, a second set of faders on each channel, and automation. All these features add to the cost of a console, but they also add to the size as well. Every time there is another feature added, more space must be allotted on the console, and analog circuitry can be big and can produce a lot of heat. The latest trend has been to make analog consoles that are packed with great features without being so big that you have to stand up and lean forward (over a "furnace") to turn a knob. Some wonderful analog consoles are available that are small but have lots of features; however, none can compare to the small size and huge features of a digital console.

A digital console is basically a computer processor with lots of faders, buttons, and knobs to control it. They have all the same features as analog consoles and more. They have reverb and effects, expandable ins and outs, and practically unlimited routing and bussing options. They have these options because it is easy to add more features; they just use more processing power. They don't need more circuits and different connections for every feature.

Many digital consoles are fully automatable and recallable. This makes them useful in live shows where different settings can be pulled up for each song or act.

One of the downfalls of digital consoles is that they do not have dedicated knobs for each channel's equalizer or sends. You usually have to select a channel and use a dedicated set of EQ knobs to adjust for each one. There is also usually a main display with menus, and many features are buried a number of windows deep. Many people do not like the sound quality of digital consoles because they don't have the "richness" of analog. Others say that the sound is clean and uncolored and perfect to work with. Most of all, digital consoles provide a lot more features for a much smaller price tag than their analog counterparts.

In the home studio, either analog or digital consoles can work wonderfully. It all depends on the purpose of the console in the studio. Is it there to just provide preamps, is it the heart of the studio, or is it just a mixing device? The answers to these questions may lead you toward an analog console, a digital console, or maybe no console at all.

Figure 6.3
A studio based around a laptop with a control surface helping the user "mix in the box." (Photo courtesy of iStockphoto, Gremlin, Image #3249041.)

Mixing in the Box or in the Console

One of the latest trends in audio is to "mix in the box." This means that all mixing functions, including equalization, compression, reverb, sends, leveling, and panning, are done inside the computer (the box) by a software DAW. The draw of this approach is that there is tons of power to be harnessed by a good computer, the software's graphical user interface is simple and easy to navigate with a mouse, and the price per track ratio is far cheaper than any console option. Some of the problems are that there aren't physical knobs to turn or faders to slide, it is not as intuitive as the flow of a console, and some believe that it doesn't sound as good. I personally feel that many of these opinions are often tainted by experience and stubbornness. Maybe it seems more difficult to mix in the box than on a console because you've always mixed on a console. Maybe it doesn't necessarily sound better mixed on a console, it just sounds different. And maybe if you want knobs to turn you can try using a control surface (see the discussion later in this chapter). Basically, if you have no or little experience on a console, there is little or no reason to use one. Mixing in the box will give you far more options and flexibility. As for functions other than mixing that come with a console, most audio interfaces have everything you need.

Audio Interface Mixer Functions

So, what are the functions found on an audio interface that defeat the need for a mixer or console? First is the microphone preamp. Traditionally preamps have been found on a large console. Now preamps come individually and built into audio interfaces. The quality of the preamp varies with the quality of the audio interface, but one thing is for sure: If you use preamps from an interface that is the similar price of a console, the preamps will be of similar quality or even a little better than those on the console. Either way, as mentioned in the previous chapter, a standalone microphone

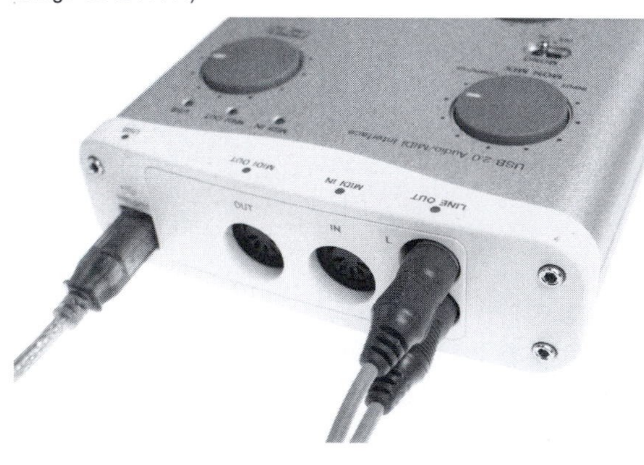

Figure 6.4
An audio interface, including a USB connection, MIDI in/out, and analog audio outputs. (Photo courtesy of iStockphoto, Marek Walica, Image #5401177.)

preamp (at a minimum of $500 per channel) will always be superior to most any audio interface's, or less than $10,000 console's, preamps.

Second is the equalizer section. This issue is also discussed in the previous chapter. EQ is really not a necessity during tracking (recording tracks). It is better to use equalization during the mix-down while you're mixing in the box.

Third are routing options. Many audio interfaces have extensive routing options, basically so that any input can feed any output. These options are usually controlled by software that comes with the interface. Even if the interface does not have these options, it is possible to route audio to and from different ins and outs via a software DAW.

Fourth is monitoring and headphone-level control. Most audio interfaces have a monitor out that is controlled by a master volume knob. When the monitor output is connected directly to powered monitors or an amplifier and speakers, the master volume knob can be used to adjust the volume of the speakers in the room. If a main output is plugged directly into powered monitors, there will be no level control and the audio will be painfully loud and might blow the speakers. One or more headphone amplifiers are found on almost every audio interface available. Every time there is a headphone connection with a volume control, there is a headphone amplifier. Most headphone amplifiers are powerful enough to drive more than one pair of monitors using a splitter cable if needed.

Fifth is talkback. Many larger consoles come with talkback functions. Talkback is the way an engineer in the control room can communicate with a performer in the studio. It is usually a microphone connected to a switch that turns on in the performer's headphones when activated. Some audio interfaces have this feature, but most do not. Many people simply plug an extra microphone into an additional input and pull up that input on a track in their DAW. They then use this microphone to talk back to the performer in the studio. Many home studios don't even need talkback because everyone is contained in one room.

Sixth is avoiding delay or latency. Latency occurs when a signal is being recorded into a DAW through the AD converter inputs and is simultaneously being monitored out the DA outputs. The problem

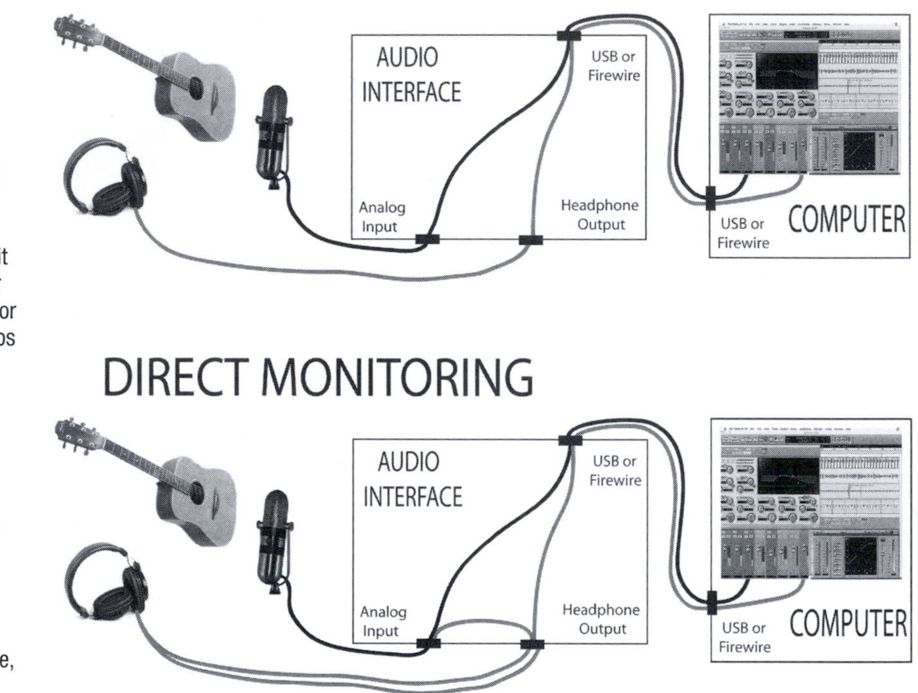

Figure 6.5
The difference between latency monitoring and direct monitoring. Latency monitoring sends the signal through the computer before returning it to the headphones. The time that it takes to go to the computer and back produces latency or delay. Direct monitoring skips going to the computer by sending the signal directly to the headphone output. This is usually added to what is playing back from the computer, and the signal in the headphones is not delayed at all. (Photos courtesy of iStockphoto, David Lentz, Image #2394240; Ben Thomas, Image #2764785; Immersive, Ben Harris.)

is that the AD/DA conversion process takes time, so by the time the signal gets back to the listener it is delayed. One of the solutions to this problem (latency monitoring) is to simply make your RAM work harder and push the information through the conversion process faster. This can be hard work for the RAM, and many computers just can't do it well enough to sufficiently minimize the latency. Another solution is to put your channel into a mixing console and simultaneously route the signal back to your speakers or headphones and into the DAW for recording. This lets the sound be monitored without going through any conversion into the DAW; therefore there is no latency. So, now you need a mixer … wait, no, you don't. Direct monitoring is an audio interface's version of this technique. Many audio interfaces utilize direct monitoring to take an incoming signal and send a portion of it to be converted and recorded into the DAW and another portion to be sent back to the listener for zero latency (or direct) monitoring.

So, basically, am I saying you shouldn't get a mixing console? No, just don't think that you *have* to get a mixing console. There are many reasons to get and use a mixing console in a recording studio. Some people who have learned to mix on a mixing console want that functionality in their studio. Other people already have a console for live sound use and want to utilize it in the studio also. Finally, there are people who want a console because it looks big and impressive. There are lots of people who keep their old consoles in their studio just for the "wow factor." The console usually isn't connected to anything or it is used very minimally, but it looks cool. My final disclaimer is that even for these legitimate reasons I have seen many people go through crazy hybrid setups to utilize the features in both a software DAW and a mixing console simultaneously. Most of the time it is a hassle and just not worth it.

CONTROL SURFACES

Control surfaces provide tactile control of onscreen functions of a software DAW. That is to basically say that they function as a glorified mouse. Most control surfaces can't really do much more than what can be done with a mouse. There are multiple levels of control surfaces that fall into two main price ranges: expensive and super-expensive.

Figure 6.6
A prosumer-level (not quite professional and not quite consumer) control surface. This is used to replace a mouse controlling software DAW functions from a hardware interface. (Photo courtesy of iStockphoto, Daniel Halvorson, Image #459000.)

The expensive controllers are the prosumer models. These units commonly have eight faders and a small master section. They usually have dedicated buttons for transport controls, various windows, and edit commands. The faders can be banked (eight at a time) or nudged (one at a time) through the virtual faders in a DAW so that all tracks can be accessed and controlled. If banking doesn't work for you, there are expansion units adding eight additional faders at a time for a similar price to the main unit. Most of these units connect to the computer via MIDI or USB transmitting MIDI. The faders are then limited to 128 steps of resolution (which can be too low for some people) and eight faders per MIDI channel. There are a few Ethernet-based controllers in this price range as well. If the host DAW supports it, these interfaces can have very quick responsiveness and excellent functionality with 1024 steps of resolution on the faders. There are also a few one-fader control surfaces meant to nudge through multiple channels. These units are rather affordable, but the one-fader design can be cumbersome to use. They also have a small master section with dedicated or assignable knobs and buttons.

The problem with the prosumer control surfaces is that they are not functional enough to completely leave the keyboard and mouse behind. Many people with these units have to remind themselves to put down the mouse and use the control surface that they paid good money for. They do look impressive and fulfill the need for the "wow factor" in many studios.

The super-expensive options are the professional models. These control surfaces are designed to function in professional full-time production facilities. They are costly but very functional, allowing the user to sometimes completely avoid the keyboard, mouse, and even the computer screen. These units come standard with anywhere from four to 24 faders. They usually have a large master section with lots of buttons, knobs, and light-emitting diodes (LEDs) for visual feedback. Digidesign calls its ICON series *worksurfaces* instead of control surfaces because they do more than just control onscreen functions. There is sufficient feedback from the console that the user doesn't have to look at the screen. Euphonix bases its professional control surface around a keyboard and mouse with a large number of assignable buttons and knobs and a master touch screen.

Some touch-screen control surfaces are also available that have great features and excellent functionality. Most professional control surfaces include monitor controls, headphone outputs, and talkback so that they can completely replace the need for an analog or digital console. The next step in control surfaces will simply be a touch-screen computer that allows you to grab faders, twist knobs, and edit waveforms with your fingertips. There will probably always be issues, arguments, and debates about what is a more comfortable or ergonomic way to control a computer interface.

Figure 6.7
Digidesign's ICON series console is used in many professional facilities. It controls on-screen functions but also gives sufficient visual feedback to mix on the console without having to look at the computer screen. (Photo courtesy of iStockphoto, Lauri Wiberg, Image #5808510.)

STUDIO MONITORS

Studio monitors are just a fancy name for speakers that are designed to be used in a recording studio. Many people wonder why they can't use just any good set of speakers in a studio, and there are a lot of reasons why not. Some of these reasons include accuracy, ear fatigue, and power.

Figure 6.8
Studio monitors are high-quality speakers designed for accurate sound reproduction to provide critical listening during mixing. (Photo courtesy of iStockphoto, Chris Scredon, Image #4078177.)

Accuracy, Power, and Fatigue

Studio monitors are designed for playback of audio in a critical listening environment. They are designed to be brutally honest, whereas regular speakers are usually designed to simply sound good. Studio monitors are detailed and accurate with a flat frequency response. This is crucial in mixing because if a speaker has too much bass, for example, you will actually turn the bass down on your mix to make it sound balanced. Then when you play your mix for someone, the bass will be weak because you compensated for the extra bass in your speakers. Regular speakers are not flat at all. They usually have boosted treble and bass to make them sound louder and punchier.

Studio monitors are built to be potentially listened to for up to 12 or 15 hours a day. They are often designed to avoid fatiguing the listener's ears. When a signal is amplified and processed poorly, it creates artifacts and distortion of harmonics (or various frequencies in the signal). When speakers get louder, those artifacts and distortion are what hurt our ears. If a signal is processed and amplified cleanly it can actually be turned up fairly loud without fatigue. If an engineer's ears are fatigued they can't make good mixing decisions. This is an aspect to consider when you're searching for monitors. We each have different aspects of sound that fatigue us or make us want to go on working forever. When you're trying out speakers, you want to find the ones that make you want to listen all day long.

The power requirements of studio monitors are different from those for regular speakers. This is best explained by looking at three types of speakers and their priorities. First, there are audiophile speakers that are designed to be very clean and accurate, but not necessarily be turned up too loud. These speakers are delicate and can be ruined if given too much signal. Next are live sound reinforcement speakers that have the main purpose of getting loud. They need to sound good, but mainly they need to get loud. Finally, there are studio monitors that need to do a combination of both of the previous functions: They need to be clean and accurate for critical mixing but need to get loud for impressing visitors and checking the low end of mixes for sufficient thumpage.

Active vs. Passive

There are two main categories of studio monitor: passive and active. Passive speakers are similar to most regular speakers. They have a speaker cable input and a passive crossover inside the speaker to send high frequencies to the tweeter and midrange and low frequencies to the woofer. A power amplifier is needed to drive the speakers, and any amp will do as long as it has sufficient power. The problem with passive speakers is that a huge part of making the speakers clean, accurate, and less fatiguing lies in the combination of the speaker and the power amp. If someone designs a great speaker and the user powers it with a horrible power amp, it won't ever sound as good as it could.

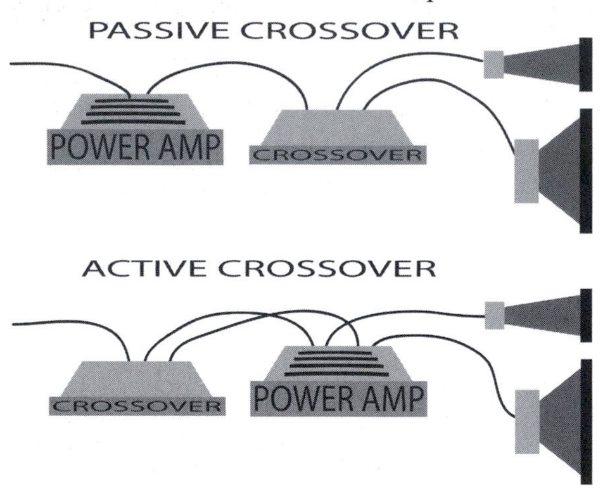

Figure 6.9
The difference between active and passive crossover networks used in and out of speakers. (Created by Ben Harris.)

This is one of the reasons that companies make active studio monitors. Active speakers have a line-level input similar to the input of a power amp. That signal is first sent to an active crossover, where high frequencies are sent to one power amp for the tweeter and midrange and low frequencies are sent to another power amp for the woofer. The benefit of this design is that each speaker can be perfectly matched with each power amp. This technique produces great-sounding speakers that are clean, accurate, and loud. The only downfall is that they are heavy because they have multiple power amps inside and they have to be provided with electrical power as well as an audio signal. Other than this they are amazing and work wonderfully everywhere in studio.

Near, Mid-, and Far Field

Most monitors that are used in home studios are near-field speakers. There are different types of studio monitors for the three monitoring positions (near field, midfield, and far field). *Near field* is usually any placement less than 5 feet from the listener; *midfield* is roughly 5 to 10 feet away, and *far field* is 10 feet or more. These types are not simply differentiated by how far away you place them from your listening position; they each have different power ratings and projection. Therefore you place them at these varying distances.

Figure 6.10
Two sets of monitors in the midfield position, placed a little behind the console. (Photo courtesy of Immersive Studios, Ben Harris.)

For example, most near-field monitors have a woofer with a diameter of 8 inches or less and usually pull 200–400 watts of power. If you place a weak and small pair of near-field monitors in the far-field position in a large room they will not be sufficiently powerful and you'll probably blow them. Midfield speakers are usually larger than near fields or will often have three drivers with an 8- to 12-inch woofer. Acoustically this is the best position in which to place your speakers—not too far and not too close.

Depending on the size of the room, a powerful set of near fields with a subwoofer could work great. Far-field monitors are large, often having two 15-inch woofers, one or two midrange drivers, and a large tweeter. These monitors have two purposes: to get loud with lots of bass and to look large and impressive. For acoustic reasons stated earlier, they usually aren't that great to mix on. They can get really loud and fill the room with sound, so they are wonderful when there are a lot of people looking to be impressed by the great-sounding mix.

Figure 6.11
A standard pair of studio headphones. Notice that they don't have on/off switches, they are not colorful and flashy, and they are definitely not earbuds. (Photo courtesy of iStockphoto, Ben Thomas, Image #2764785.)

HEADPHONES

Headphones are similar to studio monitors in many ways. They need to be clean, accurate, and powerful. They also need to be headphones that are made to be used for recording. Consumer headphones don't quite cut it. My advice is to basically avoid any headphones that you can purchase at a computer or electronics store, that have volume or bass adjustments on them, and that are called DJ headphones or anything other than for recording. If you purchase headphones from a music or recording equipment retailer and choose from some of the main brands (Sony, Sennheiser, Audio Technica, Ultrasone, etc.), you will be just fine.

There are two main uses for headphones in the studio. One is as a secondary check while mixing. The other is as a listening device during tracking. Quality headphones are necessary for both these purposes. When it comes to listening in the studio, $50–100 headphones can do just fine as long as you follow the guidelines stated earlier. This level of headphone will work fine for critical monitoring, but you could easily spend more on a nice pair of listening headphones.

Open vs. Closed

Headphones come in two major flavors: open and closed back. Open-back headphones do just what they say. The outside cover of the earpiece is not sealed so that the backside of the speaker can push into the air. The benefit of this design is that these headphones have more realistic reproduction and a more open sound. The downfall is that if you track with them, they are more likely to bleed into your signal (because of the exposed speaker driver). Closed-back headphones are basically the opposite. The earpiece has a sealed backside that completely closes off the backside of the speaker. This produces little or no leakage (even at high volumes) while tracking but makes playback sound a little less open and not as accurate as open-back headphones. Many people do not like the pressure that closed-back headphones put against their heads. It is not bad for you; it just feels different.

CHAPTER 7
Putting It All Together

The last few chapters have talked about all the pieces of equipment you'll need for your home studio, but how do all the pieces connect? What does a basic system look like, and how does everything plug into everything else? The following sections answer these questions in a few scenarios.

CONNECTION SCENARIOS

Connection Scenario 1

This scenario includes the most basic setup possible. First, the computer is a nice, simple, powerful laptop with both USB and Firewire connections. It is loaded with software from one of the five main DAWs. There is a two-channel USB audio interface that is equipped with two microphone/instrument inputs, monitor outputs with a monitor-level control knob, and a headphone output and control knob. This interface connects to the computer with a USB cable. The drivers for the interface are installed on the computer, and once the computer recognizes it, it is discoverable by the DAW program in the audio preferences. There is a USB MIDI keyboard controller connected to another USB port on the computer. Its drivers are loaded or it is automatically recognized by the operating system and subsequently the DAW in the MIDI preferences. Both of these USB devices may be powered through their USB connections or with a provided power cable. There is an external Firewire drive connected to the Firewire port on the computer with the provided Firewire cable. The drive is connected to a power strip with its provided power connection. This drive was chosen because the DAW manufacturer listed it as a good drive to use with its software program. All projects will be stored and run off this drive.

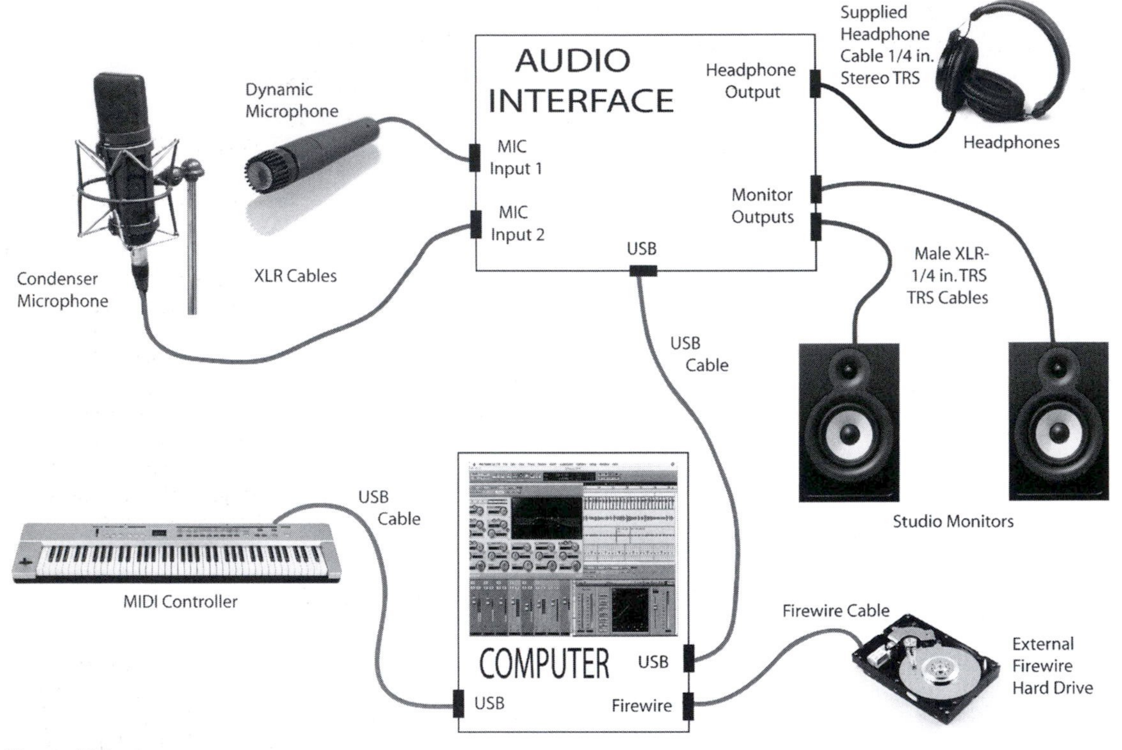

Figure 7.1
Connection Scenario 1, displaying all the physical cables and connections needed to connect each piece of equipment to the other. (Photos courtesy of iStockphoto, Ben Thomas, Image #2764785; Dave Long, Image #4431924; Mike Bentley, Image #445143; Chris Scredon, Image #4078177; Oleg Prikhodko, Image #1591800; Brandon Laufenberg, Image #6693540.)

A pair of active monitor speakers is plugged into a power strip with provided power cables, and the monitor output from the audio interface is connected to the input of the speakers with a pair of balanced line-level ¼-inch TRS to male XLR cables. The level of the speakers is now controllable by the monitor level knob provided on the audio interface. A pair of studio headphones is connected to the ¼-inch TRS stereo headphone input of the audio interface. The headphone control knob on the audio interface controls the headphone level. For a MIDI-based production studio, the setup is complete using sounds that come with the software DAW controlled by the keyboard controller. For a studio that needs to record live instruments, there are a few additional items. There is one microphone stand, an all-purpose large diaphragm condenser microphone, a dynamic microphone (a Shure SM57 or something similar), and a

15–20-foot XLR microphone cable. Either microphone can be attached to the microphone stand with the clips provided with each microphone. The microphone is connected to the female XLR connector of the microphone cable. The other end of the cable is connected to one of the microphone preamp inputs of the audio interface. If the condenser microphone is connected, it needs to be powered by phantom power. At this point everything is present and you can start making music.

Connection Scenario 2

This next scenario includes a simple yet powerful setup. The computer is a desktop computer or computer/monitor combination with USB and Firewire ports. It is loaded with software from one of the five main DAWs. There is a four-channel Firewire audio interface that

Figure 7.2
Connection Scenario 2, displaying all the physical cables and connections needed to connect each piece of equipment to the other. (Photos courtesy of iStockphoto, Ben Thomas, Image #2764785; Dave Long, Image #4431924; Mike Bentley, Image #445143; Chris Scredon, Image #4078177; Oleg Prikhodko, Image #1591800; Brandon Laufenberg, Image #6693540.)

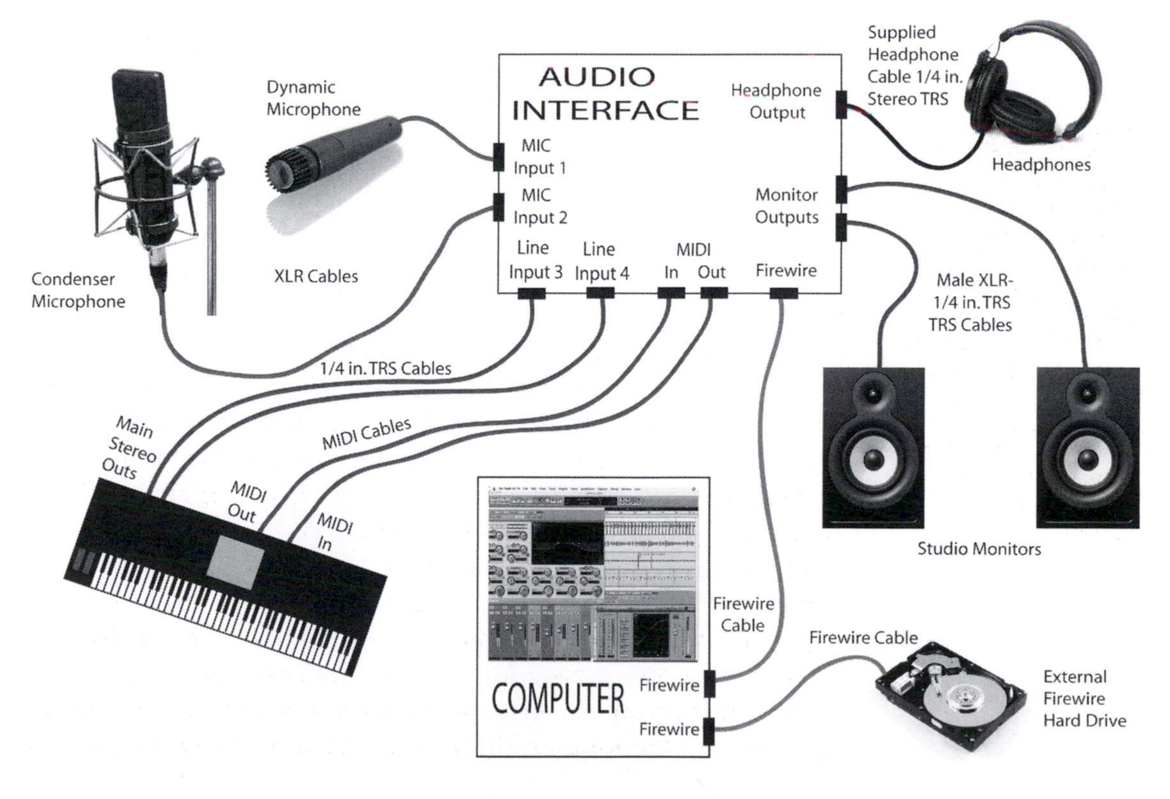

is equipped with two microphone/instrument inputs and two line-level inputs, MIDI in and out connections, monitor outputs with a monitor-level control knob, and a headphone output and control knob. This interface connects to the computer with a Firewire cable provided with it.

The drivers for the interface are installed on the computer; once the computer recognizes it, the interface is discoverable by the DAW program in the audio preferences. There is a workstation keyboard connected from its MIDI out and in ports to the MIDI in and out ports of the audio interface via two five-pin MIDI cables. These MIDI inputs are automatically recognized with the audio interface and can be chosen in the DAW's MIDI preferences. There is an external Firewire drive connected to the Firewire port on the computer with the provided Firewire cable. The drive is connected to a power strip with its provided power connection. This drive was chosen because the DAW manufacturer listed it as a good drive to use with its software program. All projects will be stored and run off this drive.

There is a pair of active monitor speakers that are plugged into a power strip with their provided power cables, and the monitor output from the audio interface is connected to the input of the speakers with a pair of balanced line-level ¼-inch TRS to male XLR cables. The level of the speakers is now controllable by the monitor-level knob provided on the audio interface. There is a pair of studio headphones connected to the ¼-inch TRS stereo headphone input of the audio interface. The headphone control knob on the audio interface controls the headphone level.

Connection Scenario 3

This final scenario includes a powerful multichannel setup. The computer is a desktop computer with USB and Firewire ports. It is loaded with software from one of the five main DAWs. There is an eight-channel Firewire audio interface that is equipped with four microphone/instrument inputs and four line-level inputs, MIDI in and out connections, monitor outputs with a monitor-level control knob, and a headphone output and control knob.

This interface connects to the computer with a Firewire cable provided with it. The drivers for the interface are installed on the computer, and once the computer recognizes it, the interface is discoverable by the DAW program in the audio preferences. There is

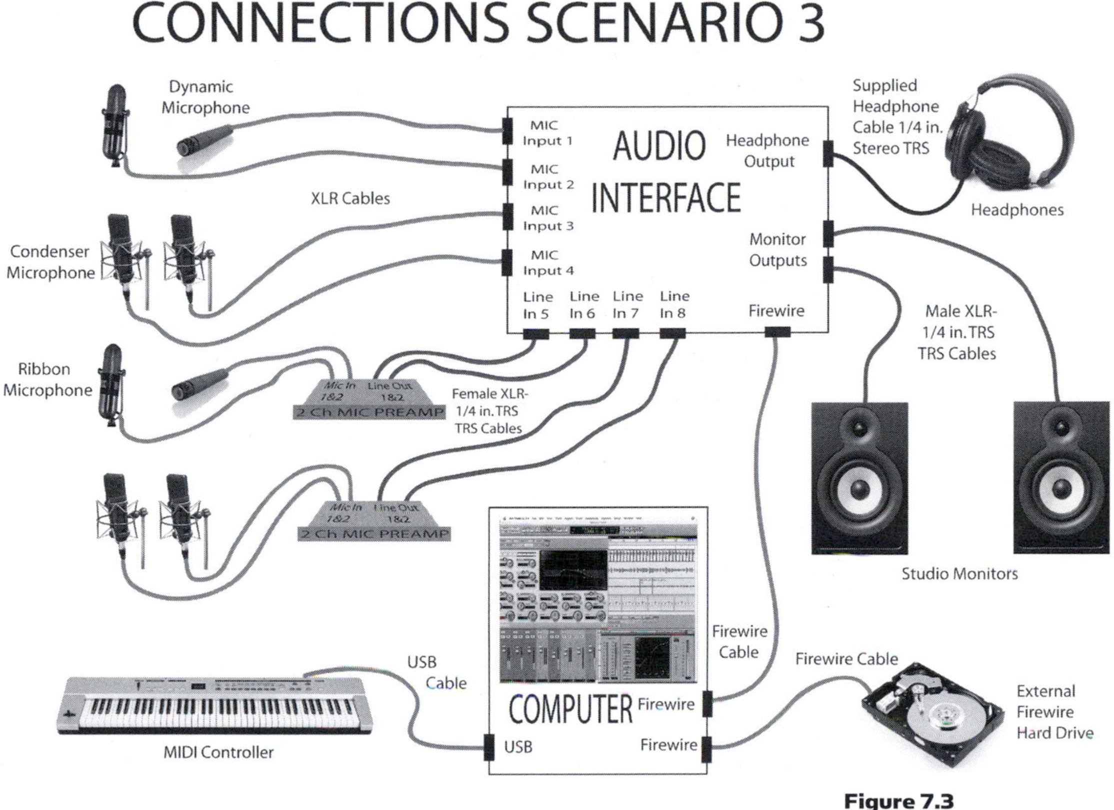

Figure 7.3
Connection Scenario 3, displaying all the physical cables and connections needed to connect each piece of equipment to the other. (Photos courtesy of iStockphoto, David Lentz, Image #2394240; Ben Thomas, Image #2764785; Dave Long, Image #4431924; Mike Bentley, Image #445143; Chris Scredon, Image #4078177; Oleg Prikhodko, Image #1591800; Brandon Laufenberg, Image #6693540.)

a USB MIDI keyboard controller connected to a USB port on the computer. Its drivers are loaded or it is automatically recognized by the operating system and subsequently the DAW in the MIDI preferences. This USB device may be powered through the USB connection or with a provided power cable. There is an external Firewire drive connected to the Firewire port on the computer with the provided Firewire cable. The drive is connected to a power strip with its provided power connection. This drive was chosen because the DAW manufacturer listed it as a good drive to use with its software program. All projects will be stored and run off this drive.

There is a pair of active monitor speakers plugged into a power strip with provided power cables, and the monitor output from the audio interface is connected to the input of the speakers with a pair of balanced line-level ¼-inch TRS to male XLR cables. The level of the speakers is now controllable by the monitor-level knob provided on the audio interface. There is a pair of studio headphones connected to the ¼-inch TRS stereo headphone input of the audio interface. The headphone control knob on the audio interface controls the headphone level. For a MIDI-based production studio, the setup is complete using sounds that come with the software DAW controlled by the keyboard controller.

FURNITURE, DÉCOR, AND ERGONOMICS

So far we have talked about the functional equipment in a studio, but isn't a studio about making music? How can you make music in a sterile uncreative environment filled with black boxes? You really can't. You need to have some sort of vibe if real and artistic music is going to be created.

Setting a Vibe

There are multiple ways to set a vibe in a studio. One starting point is to find a piece of visual art that really moves you. Find out what about it inspires you (the colors or shapes) and incorporate those elements

Figure 7.4
These photos show a studio with a great vibe. (Photos courtesy of Immersive Studios, Ben Harris.)

in the design of your studio. Some studios have a western vibe, with wood slat walls and western art. Others have a psychedelic décor, with decorative rugs and black and fluorescent walls. And some studios just have a modern vibe, looking technically futuristic yet relaxing.

Lava Lamps and Lighting

Lava lamps are a great way to set a creative ambience in a studio. They produce wonderful colors and spark wild creativity. Their only downfall is that they produce a lot of heat. Many studios turn their lava lamps off during the summer months but keep them on 24/7 during the winter. Lighting can play a big part in setting a creative mood. There are many colored light options, great-looking LEDs, and creative positioning with track and recessed lighting.

Figure 7.6
Displaying instruments on the wall in a studio can be both functional and artistic. (Photo courtesy of Foundation Studios, Ben Harris.)

Figure 7.5
Lava lamps are a mainstay in many recording studios, helping to create a creative atmosphere. (Photo courtesy of Wind Over the Earth, Ben Harris.)

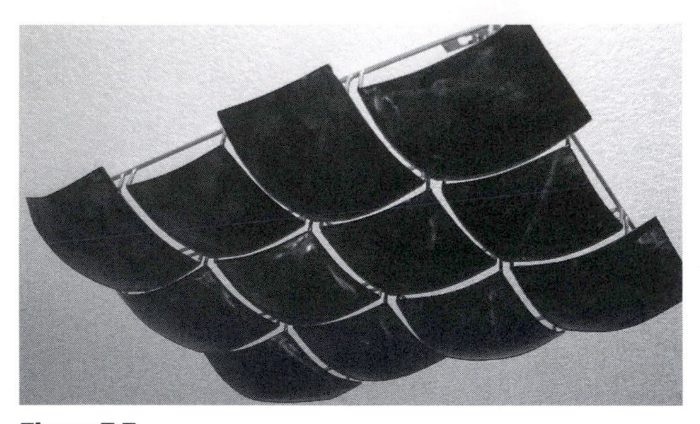

Figure 7.7
A three-dimensional decorative art piece being used for diffusion above the listening position. (Photo courtesy of Sam McGuire.)

Displaying Art and Instruments

Placing art on the walls is a great mood setter but can be tough when you have a lot of things to store and use in a studio. For this reason, displaying instruments such as guitars on the walls is very effective as an efficient storage place and a great vibe setter. A room with instruments displayed on the walls invites and excites performers to create music.

Creative Acoustics

Another multipurpose solution is to make your acoustic treatment add to your vibe. There are lots of great examples such as using three-dimensional wall art for diffusion, being creative with fabric colors over absorption, and making acoustic treatments into artistic geometric shapes.

Furniture and Décor

Furniture is another way to set the décor. New hip furniture sets a new hip vibe. Vintage and aged furniture sets a vintage and sentimental vibe. So, choose your furniture wisely to create a fitting vibe for you and your studio. Overall you want your studio to be an inviting and creative place. Use these methods and whatever you can imagine to make it happen.

Figure 7.8
The furniture in the back of a studio where the artists and clients can sit back, relax, and enjoy the music. (Photo courtesy of Wind Over the Earth, Ben Harris.)

PURCHASING SUGGESTIONS AND EQUIPMENT CONCLUSIONS

I am a firm believer that knowledge is power, and the knowledge covered in these past few chapters should give you the power to be an informed consumer. You should now understand what elements are necessary to the home studio and which elements might not be necessary for your studio.

I have avoided talking about specific models of equipment because a companion Website, theDAWstudio.com, provides up-to-date information about current equipment pricing, reviews, user complaints, and home studio issues. These last few chapters have covered almost all the equipment used in any recording studio. There are countless options available at many varying price levels. A competent studio can be built on nearly any budget. I hope that you will be able to use the knowledge base from these chapters and the current information from the Website to more effectively purchase home studio equipment that works for you.

SECTION 3
Recording Techniques

CHAPTER 8
Recording Techniques

The next three chapters discuss recording techniques that can be utilized to produce higher-quality projects in a home studio. Many of these techniques are subjective to the project, the listener, and the circumstance. Basically, there are a lot of rules, and all of them can be broken. The key to breaking the rules is understanding what you are breaking, why you are doing it, and how it might damage the project or equipment that you are working with.

This section begins with a discussion of the fundamentals of audio and recording, including analog and digital recording, signal flow, and MIDI. Then we'll talk about the individual tools that are used to capture, process, and manipulate sound. Finally, we'll bring everything together with mixing, mastering, and distribution.

FUNDAMENTALS

Transducers

The first chapter of this book covers the basics of how sound travels through air as compression waves, creating compressions and rar-

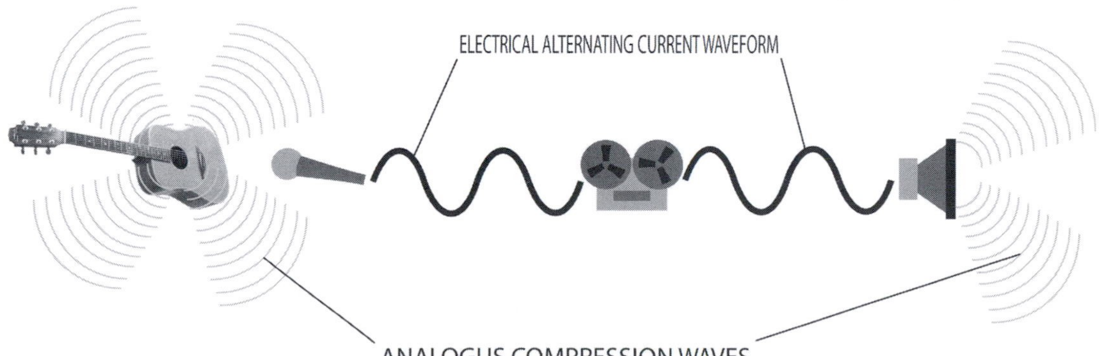

Figure 8.1
This image shows how sound in an analog system moves from start to finish. Analog signal processors can be inserted in any point of the alternating current signal. (Photo courtesy of Immersive Studios, Ben Harris/created by Ben Harris.)

efactions as it pushes air molecules back and forth. What isn't covered there is the way that acoustical energy is transformed into electrical energy so that it can be recorded and processed by analog equipment. Microphones and speakers are both transducers that convert acoustical energy into electrical energy, and vice versa. This process occurs in a similar fashion for both (mainly with dynamic microphones).

When acoustic energy varies the air pressure around the diaphragm of a dynamic microphone, it moves in conjunction with the air. This moving diaphragm is attached to a coil of wire that moves through a magnet, creating a varying magnetic field and therefore an alternating current down the wire. This alternating current is now electrical energy that is analogous to the acoustic energy captured by the microphone. If the same electrical energy is sent to a speaker, it will vary a magnetic field connected to a paper cone. The electrical variations from that alternating current will in turn move the cone back and forth, which will move air molecules, creating compression waves and therefore sound. Any piece of equipment that deals with the manipulation of this type of audio electrical energy in the form of an alternating current is called *analog*. This is because there is no additional transducer changing the source into a different form of energy. The device is simply processing the signal that is analogous to its original source.

Figure 8.2
The back rack of a recording studio filled with various signal processors. (Photo courtesy of Immersive Studios, Ben Harris.)

Signal Processing

Once sound is in the form of electrical energy, it can easily be manipulated. Frequencies can be singled out, levels can be adjusted, and signals can be added together. All this is called *signal processing*. This includes microphone preamplifiers, equalizers, compressors, limiters—all the functions of a mixing console—and amplification. Many of these functions can be done in the digital world as well as in the analog world. Sometimes a digital device cannot replicate the simplicity and beauty of an analog processor, and sometimes the limitless possibilities of a digital processor can replace the limitations of an analog unit. Circuitry and connections have a direct correlation to sound quality in analog signal processors. In digital signal processors it is the programming and algorithms used in the device that denote quality. Both analog and digital signal processors can be used to make an audio signal sound better or worse, and sometimes in mixing you need a little of both.

Figure 8.3
The first recording and playback device, the wax cylinder recorder. (Created by Ben Harris.)

Analog Recording

The first analog recording device was the wax cylinder invented by Thomas Edison. He noticed the vibrations that he felt at the end of a telephone wire as he spoke into the receiver. He realized that if he could capture those vibrations in a substance and then re-excite the phone line, he could record and play back sound. He used a cylinder coated in wax, and it worked. Wax cylinders were used through the end of World War II. In the meantime, the film world developed the optical track on motion-picture film. This format used a varying amount of light developed on the film to record and play back audio information. This technique is still used today as a backup for digital sound for film.

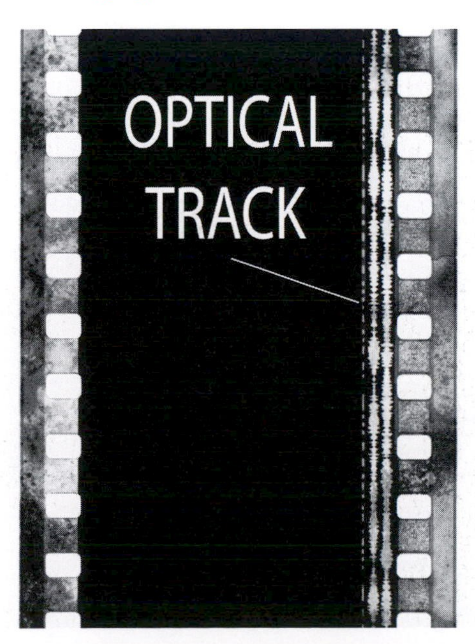

Figure 8.4
Audio recorded optically on the side of film in movies. These tracks are still recorded on modern-day films and used as a backup when the digital signal drops out. (Photo courtesy of iStockphoto, Valerie Loiseleux, Image #4928304.)

Magnet tape was developed in Germany in the 1940s, but an English soldier stole a machine and some tape from the Germans at the end of World War II. He brought the machine and tape to the United States, where the technology was reverse-engineered and produced for radio broadcasts and later recording studios. This technology captures audio information as posi-

Figure 8.5
An analog magnetic tape machine that records 24 tracks of audio on a 2-inch wide tape. This type of machine has been standard in recording studios since the 1970s. (Photo courtesy of iStockphoto, Anna Pustovaya, Image #4791935.)

tive and negative charges on magnetic tape. Also commonly referred to as *analog tape*, it is the older (and higher-quality) brother of cassette tape. Magnetic tape machines progressed through the 1950s and 1960s, placing more and more tracks on wider and wider tape. Track count eventually maxed out at 24 tracks on 2-inch tape. This is accomplished by having multiple magnetic heads (used to magnetically charge the tape or record and retrieve or play back information) that line up horizontally with different strips on the tape.

Contrary to common consumer belief and perception, analog tape sounds really good. If you went into a studio and listened to a recording straight off analog tape, you would probably be completely blown away with how good it sounds. There are two reasons people have the perception that analog tape doesn't sound good. First, they equate it with cassette tape, which is noisy and has a limited high-frequency response because of the slow tape speed. Second, advertisers have told consumers that they need "high-quality digital sound," basically badmouthing the alternative. This does carry some truth. The fact is that the delivery format of a CD or audio on a DVD is much more consistent, cleaner sounding, and closer to the original than on a cassette tape or VHS. This has no reflection on the original quality of the record medium (analog tape). As with many debates, there are benefits and disadvantages of recording to an analog or digital medium. The main considerations are that analog sounds good, but it definitely has "a sound." Analog tape is expensive and time consuming to maintain in working order. And analog tape does not give you the wonderful ease and versatility of editing and manipulation of digital audio.

Digital Recording

Recording audio on a digital medium (hard disks and optical discs) requires the information to be converted to binary code, or 1s and 0s. This is the code that all computers use to store, process, and manipulate information. Once audio has been converted to 1s and 0s, or digital information, there are near-endless possibilities of what

Figure 8.6
An analog alternating electrical current has to be converted to binary information (1s and 0s) in order to be recorded and processed digitally. (Created by Ben Harris.)

you can do to it. Two main aspects define the sound quality of your finished audio. First is the quality of the hardware AD/DA conversion. This issue was discussed earlier in Chapter 5, when talking about AD/DA converters. Basically you get what you pay for. The built-in converter in your computer is not going to sound as good as a $500 two-channel converter. And that one is not going to sound as good as a $2000 unit. And that one is not going to … stop! You can keep going up, but where do you draw the line? Feel free to consult the companion Website at theDAWstudio.com to look at these varying levels of products and help you make an informed decision as far as quality and price.

Second is the quality of the processing of digital information. Most audio-recording programs (especially the main five mentioned in Chapter 4) have a high degree of quality processing for basic record, edit, and mix functions. It is in the advanced features and plug-in processors where there are more variations in audio quality. Once again, you get what you pay for. Many of the high-quality, great-sounding processors are rather pricey as well.

Not all digital recording happens in a computer. Some of the first digital recorders functioned very similarly to analog tape machines. The difference was that they used digital tape and magnetic charges as 1s and 0s on the tape instead of positive and negative voltages. Some of these machines are still made, and many are still floating around in studios. The two main formats are the Alesis ADAT and the Tascam DTRS machines. These machines have very similar limitations to analog tape, with similar quality. The real benefit when they first came out was the price, but that is not as much an issue anymore. The current standalone digital recorders (not DAWs, as mentioned earlier; just recorders) are basically computers with a dedicated user interface for the device. They might look like a rack unit device, but inside is a computer processor and software running the recording functions. These are larger, more professional devices that are best used as replacements to analog tape machines in larger studio facilities. My suggestion is to utilize the ease of use and amazing functionality of digital recording in computer-based software DAWs.

Bit Depth and Sample Rate

Converting analog audio information into 1s and 0s is not a simple process. You can't just make all the positive voltages 1s and the negative voltages 0s. There would be nothing to define the time.

The best way to visualize this process is to compare it to a digital photograph. A digital picture is measured by pixels. Pixels are defined by a grid and placed over an image. Each pixel, or square on the grid, is analyzed and assigned a brightness or color. On the simplest level each pixel is either black or white. When there aren't very many pixels, it just looks like a bunch of black and white squares, but when you have thousands or millions of pixels you can capture a very detailed image.

Audio has another factor that is missing in a photograph: time. So, like a movie film (24 "photographs" per second), audio is sampled many times per second, with each sample capturing a view of the sound wave's current amplitude level or position. The two main

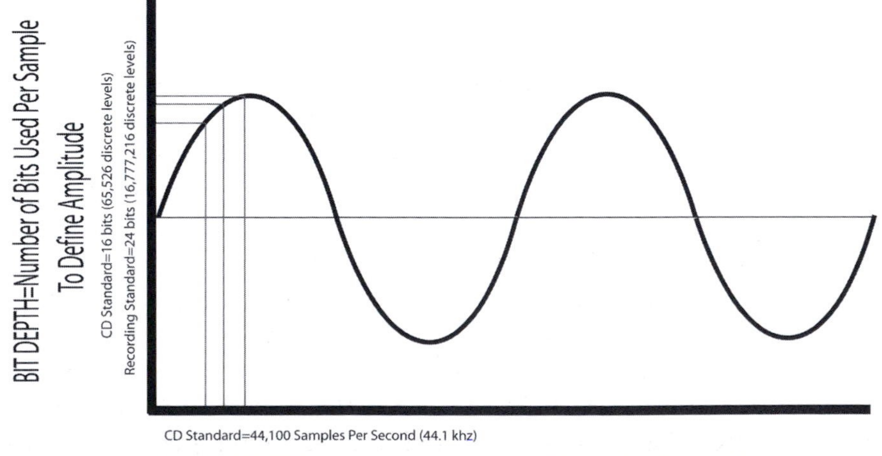

Figure 8.7
A visual representation of the analog-to-digital conversion process. As shown, each sample uses 16- or 24-bit (1s and 0s of digital information) to define its amplitude level. When all the samples are tied together and smoothed, a continuous waveform (similar to the original) is created. (Created by Ben Harris.)

defining factors of a sound wave are frequency and amplitude. The frequency of a wave is based on time, so the number of times per second the wave completes a cycle. When we're looking at a two-dimensional wave, frequency or time is defined on the horizontal or x-axis. The amplitude of a wave is determined by its pressure, the intensity of the vibrations, or how far it varies from the center resting point. Amplitude is defined on the vertical or y-axis of a two-dimensional wave. This method of creating digital audio based on defining the bit depth and sample rate is called *pulse code modulation* (PCM). WAV, AIFF, and SDII are simply PCM files. There is absolutely no sound quality difference between these formats when the same bit depth and sample rate are used. Most compressed formats, such as MP3, WMA, and AAC, are based off minimizing the amount of data in a PCM file.

The sample rate of digital audio is on the same axis of frequency and time. It is the number of times per second that a sound wave is looked at or sampled. The standard sample rate of audio on a CD is 44,100 times per second, commonly referred to as 44.1 KHz. The other standard sample rate is 48 KHz, or 48,000 samples per second, which is the standard for most film and video projects. There are higher sample rates, but they are all based on multiples of 44.1 KHz and 48 KHz. They are 88.2, 96, 176.4, and 192 KHz. The general rule of thumb on sample rate is to start with what you will end with. If you are doing a music project that will end up on CD or an MP3 format, use 44.1 KHz as your sample rate. If you are doing a project for film or video, use 48 KHz.

The bit depth of digital audio shares the y-axis with amplitude. The bit depth is a binary word (a number of 1s and 0s used to define a value). The standard bit depth of a CD is 16-bit. This means that for every sample, 16 1s and 0s are used to define a value stating the current amplitude level. Using binary code, you can define 65,536 values with 16 1s and 0s. That seems like pretty high resolution, but by adding 8 more bits to each sample (using a 24-bit word) you can define 16,777,216 values. 24-bit has become the standard for all digital audio recording due to its superior resolution and sound quality over 16-bit. Many people wonder why it is beneficial to record at 24-bit when eventually the audio has to be converted to 16-bit to be put on a CD. The answer is that when you're recording multiple tracks, processing them, and mixing them together, any lack in quality will be accentuated. By maintaining the highest quality level until the final step, you can make the degradation to your audio signal as minimal as possible.

Another answer to this question pertains to dynamic range. Since bit depth involves defining the amplitude of any given point in a waveform, it has direct correlation to the dynamic range of a signal. Dynamic range is usually measured by the number of decibels of level that are available between clipping (on the top) and noise (on the bottom). The general rule of thumb with a digital system is that if you equate 6 decibels to every bit, you get the potential dynamic range of a signal. Therefore, a 16-bit signal has 96 dB of dynamic range, and a 24-bit system has 144 dB. This doesn't mean that each additional bit defines a step of 6 dB of level; this is just a way of measuring the potential dynamic range and subsequently the noise floor of a system.

The difference between 16-bit and 24-bit is noticeable to even the untrained ear, but higher sample rates (96 and 192 KHz) are a different story. This is an ongoing and heated debate. There have been many blind listening tests attempting to prove that higher sample rates are better, but none has been able to get consistent positive results. Many of these tests have not been run properly, showing more the flaws in converters than the actual sound quality difference of the sample rate. Many engineers swear by higher sample rates, and still others use 44.1 and 48 KHz. Both produce professional quality projects, so don't feel that you have to use high sample rates to sound good.

One of the main issues for and against higher sample rates is based on the Nyquist theorem. The Nyquist Theorem states that the sample rate must be twice the rate of the highest frequency desired to be produced by the system. Since we can only hear from 20 Hz to 20 KHz, we need a sample rate of at least 40 KHz to produce the 20 KHz frequency. Think about it this way: If there is a sample on top of a wave (the compression) and one at the bottom (the rarefaction), when we take the wave away and connect the dots, we get something similar to what we started out with. Therefore we need at

Figure 8.8
A visual demonstration of the Nyquist Theorem, or essentially the reason any wave requires a minimum of two samples to be recreated. (Created by Ben Harris.)

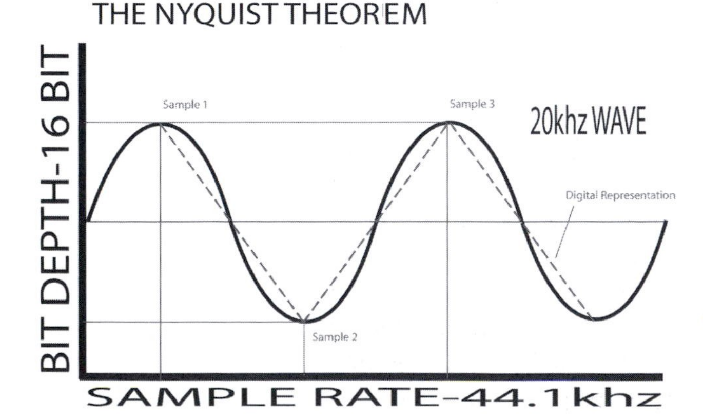

least two samples per wave to define it as moving up and down. So, if we now sample at 96 KHz, we can produce up to 48 KHz … woo hoo! Now our dogs can hear it. But there are studies that say that we can feel those upper frequencies when they occur. However, how many speakers produce over 20 KHz? Not many … and they are very expensive. So the debate rages on with arguments of detail, clarity, and smoothness pushing it onward. Check the companion Website at theDAWstudio.com for continuing arguments, advancements, and listening examples pertaining to this issue.

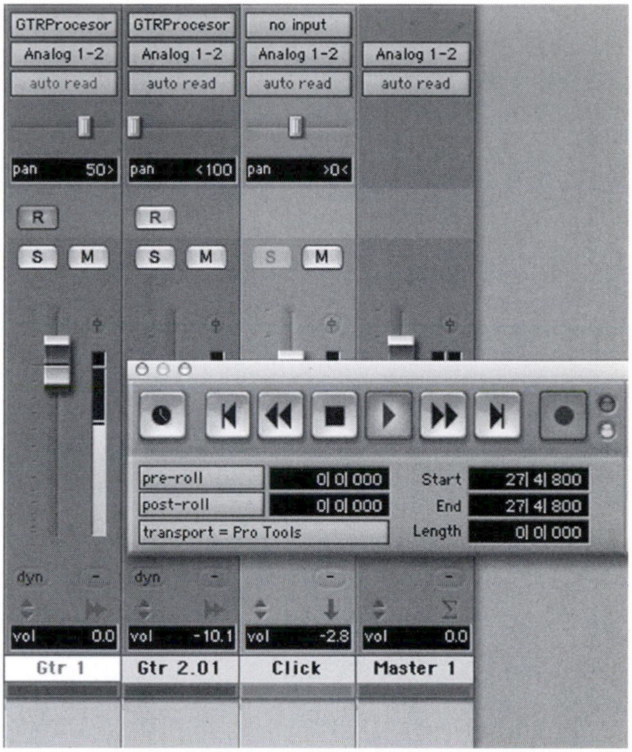

Figure 8.9
A track record enabled with the software DAW put in Record mode by pressing Record and Play in the transport controls. Notice the input level of the enabled track reaching a solid 2/3 level.

Digital Recording Levels

There is a myth that is perpetuated by forums and magazines that plagues the home studio world. That is that you need to record digital audio as "hot" as possible (or as close to clipping as possible) to maintain the highest resolution. Theoretically this is true, but realistically it causes more clipping, distortion, and overcompressed tracks than high-quality audio.

Personally, straddling the line between the professional and home studio worlds, I have asked countless people on both sides about this issue. Professionals consistently respond with a vacant look of "That is completely ridiculous," whereas home studio users testify to the myth's truth. I don't know why this myth has continued, but the truth is that you should aim for a consistent peak around 2/3 of the way up your meter during tracking. This gives you a healthy and strong signal with little fear of clipping. It also gives you a good level signal into your processors without worrying about clipping as soon as some gain from compression or equalization is added. The old myth is a valid issue when you're dealing with analog tape because of the high noise floor, but digital audio has a much larger dynamic range, so you have to record at pretty low levels to experience any loss of quality or added noise. This myth is also perpetuated by saying that professionals have hot tracks. No, professionals have good-sounding tracks and that often means that they sound louder or appear hotter, but it is not because they just turned up the volume. All the things that we have talked about—using high-quality pre-amps, converters, microphones, good-quality acoustics, and proper signal levels—are what make their tracks sound better and hotter.

Tracking

Tracking is the process of recording live tracks into a recording device. There are some basic things that need to be set to make tracking run smoothly and easily. First, most DAWs follow the same convention of record enabling, making record ready, or arming tracks. This is usually shown as a button on the track labeled *REC* or just *R*, which turns red when pressed. When these red lights are on, you are not yet recording. This is just the way that you determine which tracks are ready for recording. There is usually a main record button on the transport (the controls for start, stopping, and moving around in a session) next to the Play, Stop, Fast Forward, and Rewind buttons. Most recorders require you to first hit record and then play. At this point recording will begin on the track(s) that you have record enabled.

Most programs have key commands to start recording (basically pressing Play and Record simultaneously), such as the *R* key, the * key, or the 3 key. Find out what the key command is in your DAW and use it. It makes tracking quick and easy. One key command that is standard in most all DAWs: The Spacebar key functions as both Play and Stop, so once you are done recording, hit the Spacebar to stop.

The previous section talks about setting levels. Once you have record-enabled a track, the track's meter will show the current level of that input (if it shows no level you need to make sure that the input of that track matches the input that your signal chain is connected to). You can adjust this level by changing your analog levels of the microphone preamp or instrument input. Adjusting the fader in your DAW will not change the input level. It has an effect on the output of the track, whereas the level of the signal going into your AD converter adjusts the input level of the track. Remember to set the level of your microphone preamp or instrument input so that your track meter peaks around 2/3.

MIDI

MIDI stands for Musical Instrument Digital Interface, but what does that really mean? MIDI is a protocol transmitting four basic pieces of information:

- When a note is played
- Which note is played
- How hard a note is played (velocity, not volume)
- How long a note is played (or when a note is released)

MIDI does not carry any audio information. It simply translates human commands of pressing keys on a keyboard or pads on a drum set into these four commands. The beauty of MIDI is what is done with this information. First, any parameter of these four commands can be easily changed and manipulated. So, if you play a wrong note, it can be changed. If you play at the wrong time, it can be changed. Or if you play too soft or hard, it can be changed. Second, this information can be used to trigger or play any sound you have available to you. These parameters can command the sound of a piano and

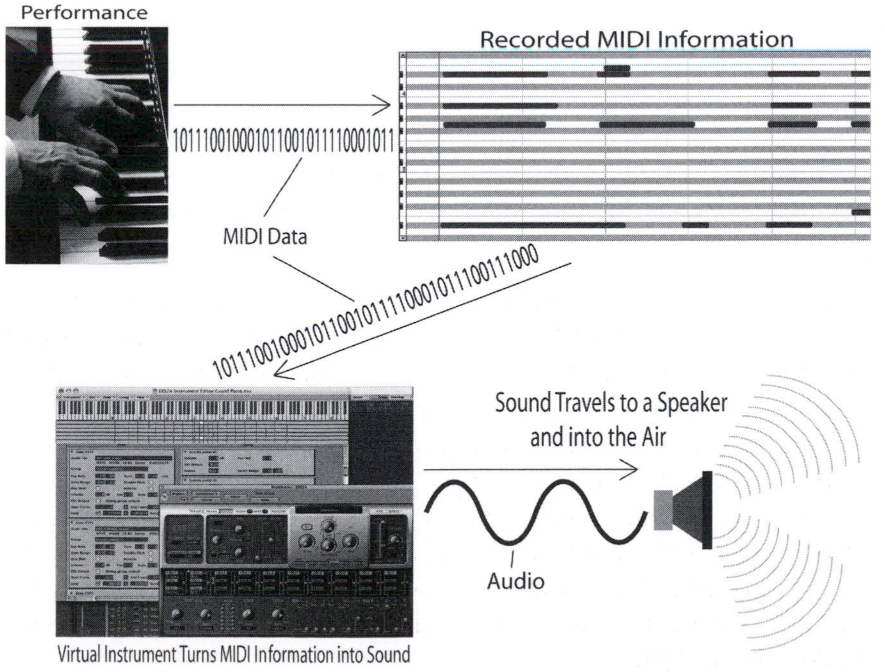

Figure 8.10
This diagram shows the signal flow of a MIDI signal. It begins with a MIDI controller being played; the MIDI information is recorded, that information is sent to a virtual instrument, where it is turned into audio, and it is subsequently sent to a speaker and then the air. (Photo of hands on piano courtesy of iStockphoto, Robert Rushton, Image #2222173.)

then be changed to a marimba and then to a drum set. The device receiving the MIDI information and assigning it to a sound produces the audio in a MIDI signal chain. This device is integrated in any MIDI keyboard that produces sound internally. It is called a *sound module* in the hardware world and a *virtual instrument* in the software world. With any of these devices, a player can use a MIDI controller keyboard (or any other MIDI controller) to command any sound available in the connected device.

The Purpose of MIDI

The purpose of MIDI is to let someone with a MIDI controller (keyboard, guitar, or drums) produce the sound of any instrument they want without having to learn how to physically play that instrument. By assigning the series of four commands we mentioned to different sounds, you could use a keyboard to make the sound of a saxophone, a trumpet, or an entire string section. This idea was blown out of proportion in the 1980s, when MIDI was first developed. Film and television music became overrun by MIDI productions. It was much cheaper and faster than hiring real musicians for every instrument. Studio musicians had to pursue different careers; listeners gained a loathing of the sound of 1980s MIDI productions, and the fad eventually wore out.

MIDI has been making a slow and steady comeback since the 1990s, but there are still many who can't get over the terrifying memories of those 1980s' soundtracks. The fact is that the sound is based on the quality of the samplers and synthesizers receiving the MIDI information. Samplers sound much better than they did in the 1980s because of the marvelous advancements of computers. Most of these sounds are so good that, if they're recorded and manipulated properly, they can fool most people into thinking that they are listening to a real instrument.

MIDI Techniques

There are two main paths to take when sequencing or recording MIDI. One is to use the tools to create whatever crazy sound, impossible performance, or mechanically precise rhythm possible. The other is to try to make the instruments sound as though they were truly performed by that instrument and not a MIDI keyboard.

The first step is to think like the instrument. How does a violin generate sound? How often does a saxophone player need to pause and breathe? What is the range of a viola? How many notes can a bass guitar play at one time, and how? Once you know the answers to these questions, you can work on performing the part as it would be performed by the real instrument. This includes practicing your performance on your MIDI controller to eventually have a real performance like the instrument. If your performance is close to the final desired product, you incorporate inconsistencies, randomness, and imperfections, which all make your track sound like it was played by a human. Another thing that you can do is to not play every part like a keyboard player. Strings don't play in chords; they each have individual lines that create chords when added together.

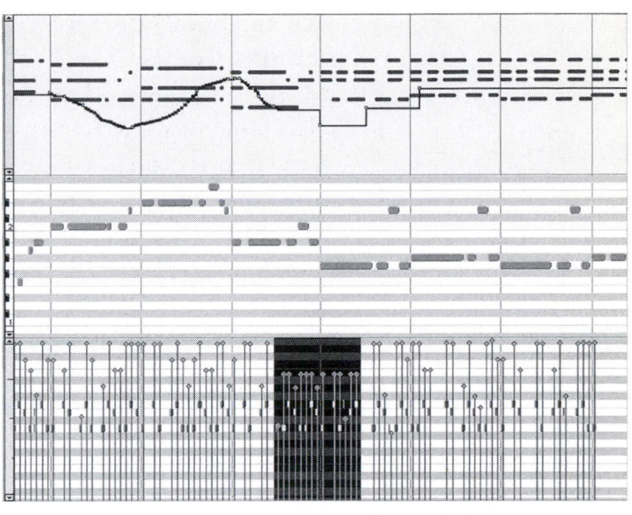

Figure 8.11
Three MIDI tracks, each with different parameters being edited. The top track displays volume, the middle track displays notes, and the bottom track displays velocity. Each of these parameters can be edited individually, making it possible to tweak almost any aspect of a MIDI recording.

There are many more MIDI parameters than the main four, and they can be used to create a better performance. The sustain pedal is one of these. Use the sustain pedal with instruments that have one in real life, such as piano, pedal steel guitar, and harp. Strings do not have a sustain pedal! The modulation wheel is another wonderful tool included on most MIDI controllers. It produces a vibrato effect by varying the pitch of the signal at a consistent rate. Not all instruments can produce vibrato. In real life they may do it in different ways and at different times during a note. Many instruments, including voice, produce a vibrato effect when holding out long notes. This can be performed with a MIDI controller by first hitting the note and then slowly bringing up the modulation wheel while the note sustains. The pitch wheel is another excellent tool for adding realism. Once again, not all instruments can do pitch bends. You must ask yourself how the instrument performs the pitch bend and how far it can bend. Once you know the answers to these questions, you can practice performing these pitch bends with the pitch wheel during your realistic MIDI performance.

Now that your performance is as close to a real performance as possible with mistakes, inconsistencies, and randomness, you have to be prepared to individually edit every note to make it right. You have to become a tweaker, ready and willing to change individual velocities, draw new pitch wheel curves, and slide notes into the correct rhythmic feel. One of the greatest and worst tools in MIDI production is quantizing. Quantizing basically moves your notes around so that they fall on a rhythmic grid (playing with a click track or to the rhythm of the session is key to making this work). This can be great if you can't quite get the rhythm right, but it can also suck all the life out of a performance, making it sound like a robot played it. The ideal place is to aim for having such a good performance that it does not need to be quantized, except for maybe drums and percussion. If you do have to quantize, lower the strength so that it maintains some of the original feel of the performance.

These techniques coupled with basic MIDI editing of pitches, velocities, and note timing can help anyone make their MIDI productions sound more like real instruments. Or they can help you make your crazy synthesizer parts and tight drumbeats sound even crazier and tighter. MIDI contains a world of possibilities if you are simply willing to go out and grab it.

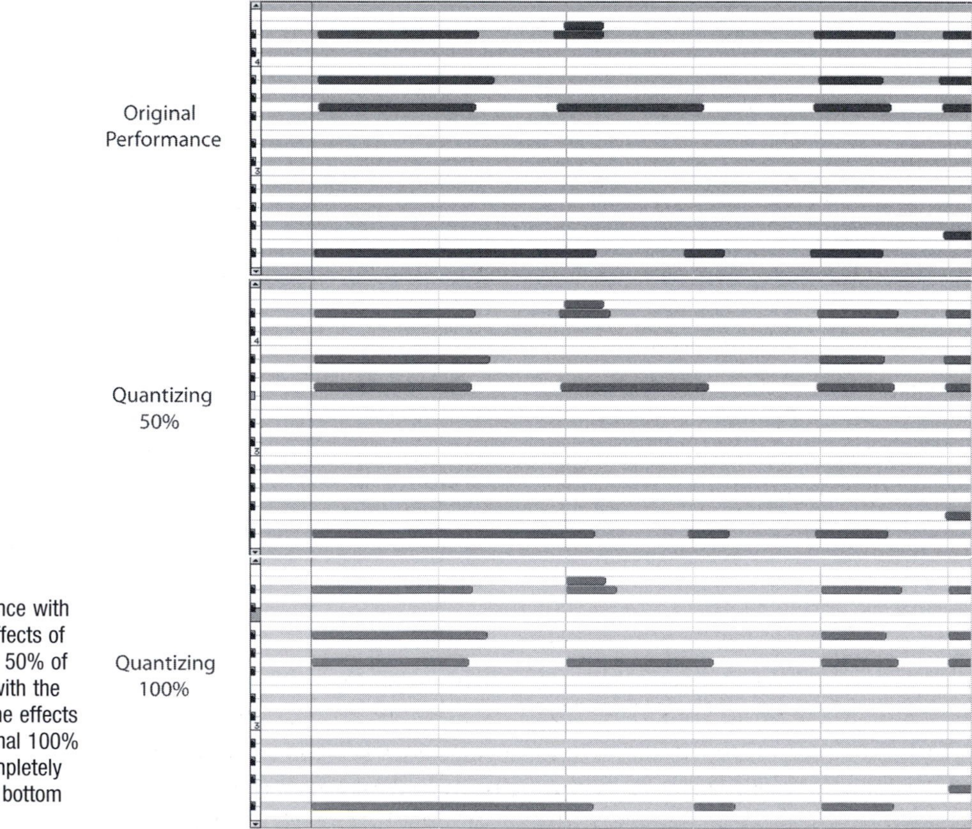

Figure 8.12
An original performance with the top image, the effects of quantizing the signal 50% of the way to the grid with the middle image, and the effects of quantizing the signal 100% of the way to (or completely on) the grid with the bottom image.

Other Uses of MIDI

The main controls we mentioned are not the only signals sent in the MIDI protocol. There is timing information (such as bars and beats and minutes and seconds), continuous controllers (pitch wheel and modulation), and system configuration commands (such as patch and program changes). A lot of this information is good to understand when working with hardware MIDI devices, but there are many devices that might not seem like MIDI devices that use these commands to transmit and synchronize information. MIDI time code (MTC) can be sent to synchronize timing between devices or software. Many control surfaces use the MIDI protocol to let you move faders, start and stop playback, and do various other functions in a DAW. Many keyboard controllers come with additional faders and knobs that can be assigned to various functions in a software DAW. The MIDI protocol is very simple and easy to work with, so there are many customized setups that use this protocol to control anything from lighting to pyrotechnics.

PRODUCTION BASICS

This book is not focused on covering music production and all its facets, but a small amount of that topic needs to be covered to get you making music on the right track. The term *production* means a lot of things. In the world of MIDI, it encompasses the entire process, from writing to recording. In audio production it is not just the recording process, it is writing, tracking, and mixing. The purpose of production is to present a song to the audience in a way that it will effectively convey the feeling, passion, and emotion of the artist. This is not as simple as putting microphones up in front of a band. There are many things that you can do to help convey the true performance of an artist. One technique is adding additional elements that complement and accentuate the intensity of the music. Many of these elements or effects can be MIDI based and benefit from having a set tempo.

Using a Click Track

If an entire production is recorded to a click track or metronome, you can utilize many of the production benefits of a DAW to add effects and elements in complete tempo, groove, and synchronization with the song. There are wonderful debates on this issue, as with any issue. Many say that recording with a click track destroys the groove of a song. This can be true with some performers, but what many are calling *groove* often ends up sounding like bad rhythm to everyone else.

If you listen to recordings of good bands with a metronome going, you'll often notice that even if they didn't record with a click track, the drummer is so good and steady with the rhythm that he or she is basically a human metronome. The art of keeping a steady tempo is rushing it or laying back on it. A steady tempo is very pleasing to the listener, and even more pleasing is the ebb and flow of the more melodic elements of the music slightly pulling away and coming back to that set tempo. If this is done well, many critics will think that the music was not recorded to a click because it doesn't sound robotic. There are some styles of music that are supposed to sound robotic, and the best way to accomplish this is to record with a click. And there are some forms of music that would sound horribly rigid if recorded with a click.

A click track or metronome is a steady pulse that gives the musician a guide to play with that follows the tempo of the song. Most DAWs begin with a default tempo of 120 beats per minute. This can easily be changed, and the click track will reflect that change. Some programs start with the click track sounding every time that you record; others require that you turn on or enable some type of metronome button to make the click track audible. You should push yourself to try to get good recordings with a click track because of its many benefits. For example, let's say that you have this great guitar riff, so you record it; but what happens when you try to record some drums with it and realize that your tempo drifted and now your drums make the drifting sound really annoying and obvious. Let's back up and originally record that guitar riff with a click track. Now when you add drums, you can quantize your MIDI drums or even use a loop and everything will sound rock solid.

Using Loops

I've noticed that people often have a hard time playing with a click track; sometimes an easy solution is to use a loop instead. Most loops will easily conform to your project's tempo so that you can bring in a drum or percussion loop that plays perfectly with the metronome. Then you can turn off the metronome or click track and let the musician track his or her performance to that loop. In the end you have accomplished the same thing as playing with a click track, but maybe it was a little easier for the musician to keep in the groove with a nice drum loop.

Loops can also be used to create a quick sketchpad for a song. You can quickly use drum, bass, and guitar loops to lay down an idea and then go back and replace some of those parts as the idea evolves. Some DAW programs have better tools than others for working with loops, but all allow you to do amazing things.

SIGNAL FLOW AND ROUTING

Signal flow is the path that an audio signal takes as it moves through analog circuits or digital software. Routing is manipulating and changing that path. Understanding signal flow is key to manipulating sound for your benefit. Audio signals flow in a similar manner to water flowing through plumbing pipes. If too much audio signal is sent through a path, the signal will become distorted, chopping or squishing the peaks of the waveform. This is similar to water being shoved through a pipe with too much pressure so that it becomes white, frothy, and aerated. Signals do not become distorted by casually moving down a path. Distortion usually occurs when a signal is amplified (sent through a larger pipe or path) and then forced into an attenuated (pushed into a smaller pipe) signal path. This can happen in any point of an audio signal.

Understanding the flow of audio and minimizing distortion at every point in the signal chain is key to having better-sounding projects. Going back to the water example, if you have a large pipe that progressively gets smaller, the final output will be distorted. If you turn the pipe around and slowly get larger as it flows, the output will be smooth and clear and very similar to the original.

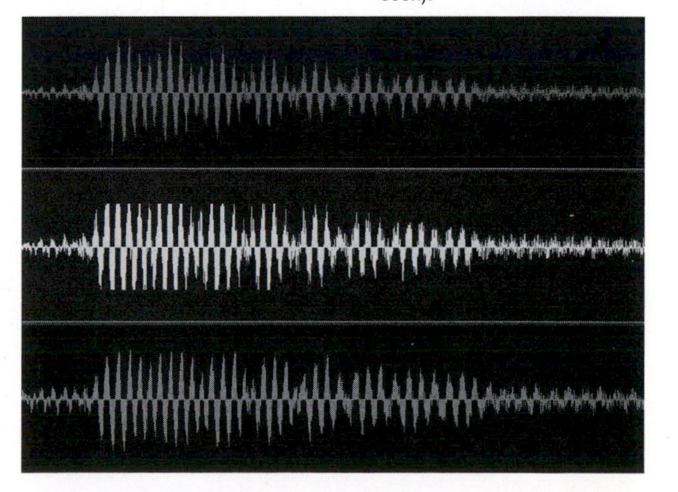

Figure 8.13
Two types of distortion: digital clipping and analog distortion. The top image displays the original waveform, the middle image displays the effects of digital clipping (basically chopping off the waveform at the clipping point), and the bottom image displays the effects of analog distortion (the signal was sent through an analog guitar processor where a general rounding off of the waveforms can be seen).

Inserts

Two main methods are used to route a signal from a track inside a mixer to a processor: inserts and auxiliary sends and returns. An insert basically inserts the processor in the flow of the signal chain. It does this by stopping the audio in its path, routing it to the processor,

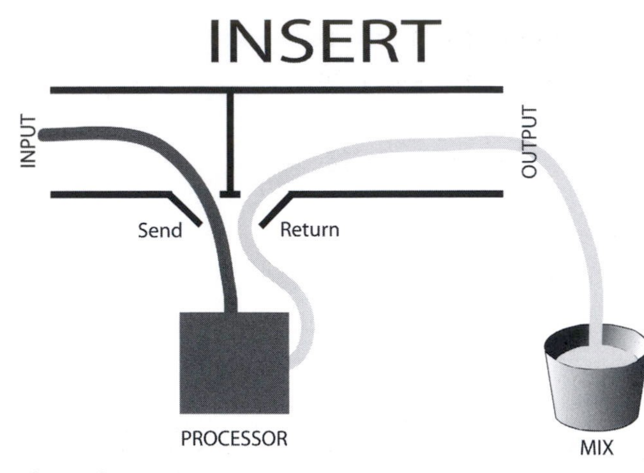

Figure 8.14
The path that a signal in a track takes when routed to an effect using an insert. Notice that once the signal is sent to the processor, no unprocessed (dark) signal makes it to the main mix. (Created by Ben Harris.)

processing the signal, returning it to its path immediately after it was stopped, and then continuing on as a completely processed signal. At this point the entire signal has been processed (wet) and there is no unprocessed (dry) signal left in the track. Inserts are usually used with equalizers and dynamics processors (such as compression and gating) because the complete signal needs to end up processed. Inserts are used to apply the majority of processors but can only affect one track at a time.

Auxiliary Sends and Returns

An auxiliary send and return does not stop the signal but simply sends a portion of the signal to the processor. The processed signal is then returned to the main mix, blending the processed (dry) signal from the track with the unprocessed (wet) signal from the processor. Auxiliary sends and returns are commonly used with reverbs, delays, and some time-based effects because each of these processors requires both wet and dry signals to be mixed together. The biggest benefit of using auxiliary sends and returns is that you can send multiple tracks to the same processor. This is how you can send multiple elements of your mix to one reverb. Each track's amount of reverb is adjusted individually from each track's send level. Most songs have one to three reverbs that are used by different tracks in the mix through auxiliary sends and returns. If a plug-in reverb were to be used on each track of a mix, the session would quickly come to a grinding halt due to a lack of power. This is why using auxiliary sends and returns is an invaluable resource when mixing in any medium.

Figure 8.15
The path that signal in a track takes when routed to a processor using an auxiliary send and return. Notice that the mix has both unprocessed (dark) and processed (light) signal. (Created by Ben Harris.)

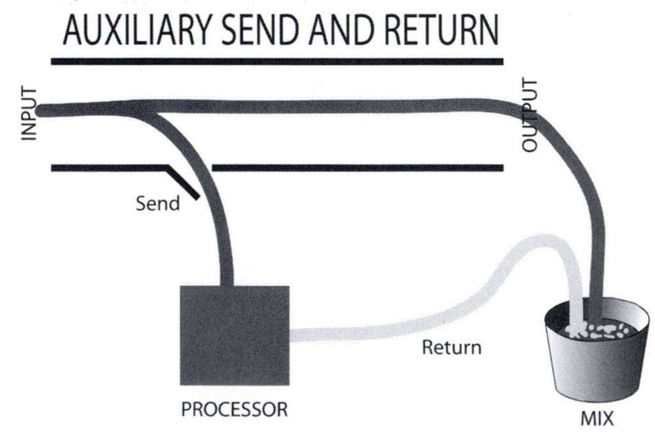

Auxiliary sends do not have to be used only to send signal to effects processors. Auxiliary sends are often used on consoles to create headphone (or monitor) mixes for the musicians performing in the studio. It works fairly simply, since an auxiliary send is sending a portion of the signal to a buss or output, you can simply connect that output to a headphone amplifier and connect the performer's headphones.

There are two modes for auxiliary sends: post-fader or pre-fader. Post-fader is usually the default and basically takes the signal after it has been adjusted by the level of the fader, so if the fader is turned up, more signal goes to the send, and vice versa. Pre-fader is often a choice by pressing a button labeled *PRE* or just *P*. Pre-fader sends the signal before it passes through the fader, so no movement of the fader has an effect on the level of the signal going out the send. Pre-fader is most often used for headphone mixes because you can create a completely different mix than the mix set on the faders. The beauty of this is that you can use multiple sends to create multiple headphone mixes so that each performer is hearing exactly what he or she wants.

Figure 8.16
The dark faders control the busses on this analog console. All tracks have to be sent to one or two of these busses before going to the speakers or any grouped output. (Photo courtesy of iStockphoto, Bill Manning, Image #5510994.)

Busses

Busses are paths on which an audio signal can travel. They are ways of taking audio from one point to another. Once an audio signal has gone through every function of an individual track, it has to be sent somewhere to eventually be heard. Many consoles have multiple busses with 1 and 2 being the main output. So, to hear a signal, it has to be sent out of the track to the main buss or busses 1 and 2 to eventually go out the console's main outputs. Other busses can be used to create submixes or to route individual tracks. Most software DAWs use busses to route the signal anywhere you can imagine inside a software mixer.

CHAPTER 9
Tools

There are many tools to be utilized in a recording studio, but the coolest ones are microphones and processors. This chapter covers microphone choice and techniques that will allow you to effectively capture any sound source. It also covers techniques to use processors to make your tracks sound better, worse, or just plain crazy. Feel free to consult the companion Website at theDAWstudio.com for reviews and pricing on current microphones and processors to assist you in your purchasing decisions.

MICROPHONES

To review, three main types of microphones are used in recording studios. First, the dynamic microphone is better on explosive sounds such as drums and electric guitars because it takes some inertia to get the mass of the diaphragm moving. Second, the ribbon microphone is delicate and very smooth sounding, with a more realistic reproduction of the way our ears hear elements. Third, the condenser microphone is very detailed, with a hyped high end, making it excellent on vocals and any instrument for which you're looking to reproduce a larger-than-life sound.

Figure 9.1 The three main types of microphones used in recording studios: ribbon, dynamic (moving coil), and condenser. (Photo courtesy of iStockphoto, David Lentz, Image #2394240; Dave Long, Image #4431924; Mike Bentley, Image #445143.)

Microphone Technique

Knowing how a microphone sounds on various instruments and in different acoustic environments is the first step to developing good microphone technique. Each circumstance requires special considerations as to which microphone should be used and where.

Figure 9.2
One common microphone placement used for acoustic guitar. (Courtesy of the University of Colorado, Denver, Ben Harris.)

Having good microphone technique doesn't mean that you know where the mic has to go in every circumstance on the first try. It means that you have a good idea where to start experimenting, and when things sound weird, you know where to try moving the mic to make them sound better. Even where engineers are using the same mics and same room every day, they have a pretty good idea where to start, but they usually have to make changes based on the instrument and playing style.

Microphone technique is one of the many areas of recording in which an artistic side comes into play. But just as a painter needs to have knowledge, technique, and honed physical skills before artistically slopping paint on a canvas, a sound engineer needs a good understanding of room acoustics, microphone patterns and characteristics, and individual instrument acoustics to make artistically guided microphone placement decisions.

Choice Considerations

Similar to a painter, engineers have a palette of colors to choose from. A lot of coloration and processing can be done in the mixing process, but capturing different sound sources in a mix with different microphones gives a song clarity, depth, and distinction. Capturing every element of a mix with the exact same microphone can be similar to painting a black-and-white image; it can be artistically beautiful, but it might miss the vibrancy and beauty of a piece full of color and life.

Each microphone has certain characteristics that make it sound unique from other microphones. Every microphone has an inherent frequency response that is derived from the type of microphone, the size and design of the diaphragm, and the electronics in the signal path. Very few microphones have a completely flat frequency response. In fact, those that do sound very boring and unmusical and are basically used for measurement purposes. Some of the most popular microphones in the history of recording are those that have an extreme change on the signal. Many of the old tube microphones from the 1950s and 1960s make a vocalist sound much better than they do in real life. There are also microphones that make a signal sound grainy and distorted for some great lo-fi (low-fidelity) effects.

Many home studios do not have the luxury of buying an extensive microphone cabinet so that they can have a different microphone for any circumstance. Many are caught in the reality of only being

Microphone Choice Consideration Chart

Microphone Type	Description	Commonly Used For:
Dynamic (moving coil)	These are rugged more durable microphones that take a little more inertia to get the diaphragm moving. Dynamics are usually a cardioid pattern microphone receiving information from the front while rejecting content from the back and sides.	Live Sound Individual Drums Distorted Electric Guitars Explosive Sound Effects Rock Vocals DJ and Broadcast Announcers Percussion
Ribbon	These are delicate microphones that reproduce a sound source more similar to how our human ear hears it. Ribbons are naturally a figure of eight pattern so they pick up sound evenly from the front and back of the microphone.	Brass and Wind Instruments Room Microphones Drum Overheads Ensemble or Distant Mics Electric and Acoustic Guitars Vocals
Large Diaphragm Condenser	These are the most common studio microphones. The large size of the diaphragm produces a resonant frequency bump somewhere between 10 and 20 khz that gives these microphones a hyped high-frequency response making them sound larger than life. Large condenser microphones can be designed with two diaphragms in order to change between multiple polar patterns.	Vocals Acoustic Guitars Drum Overheads and Cymbals Voiceover Piano
Small Diaphragm Condenser	These are also very common in the studio, but because of the small size of the diaphragm the resonant frequency is higher than the human range of hearing. This makes them have a less hyped frequency bump in the high end. Small diaphragm condenser microphones can be made for any polar pattern, but often have switchable capsules to accomplish multiple patterns with one mic body.	Acoustic Guitars Drum Overheads and Cymbals Individual Drums Piano Room Microphones Ensemble or Distant Mics Percussion

Figure 9.3
(Created by Ben Harris.)

able to afford one or maybe two microphones. The best thing to do in a circumstance like this is to find one good all-purpose condenser microphone that sounds good on everything. Most microphones over the $1000 mark are usually very specialized and sound good on only one or two types of recording source (for example, female vocal, acoustic guitar). The microphones under $1000 are more all-purpose. I would also suggest not getting a tube mic for your all-purpose microphone. These mics are often colored, sounding good on some sources but not others. A good large diaphragm condenser will work wonders for recording all types of sources.

Once you have the condenser, I would suggest one more microphone: a less-than-$100 Shure SM57, which, as we mentioned earlier, is the go-to mic for micing guitar amps and snare drums. With these two mics you should be able to sufficiently record many sound sources.

Instrument Acoustics

In the first section of this book we discussed room acoustics and the way a sound interacts in any given space. That knowledge will help you make microphone placement decisions (avoiding direct reflections and placement in corners). But another aspect that is crucial to good placement is understanding the basic acoustics of various instruments. Every acoustic instrument has a natural acoustic characteristic, or a way in which its sound travels away from the instrument. These characteristics vary based on how an instrument produces sound, whether it has a resonator, and whether it produces sound from multiple locations.

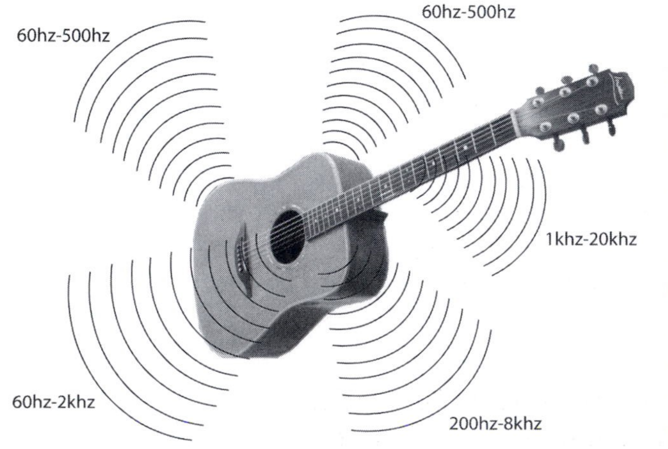

Figure 9.4
This image shows how instruments send various frequencies off in different directions. Excellent microphone placement involves having a good idea about how different instruments send off their frequencies and knowing where to put the microphone to capture the most balanced or useful sound of that instrument. (Photo courtesy of Immersive Studios, Ben Harris/created by Ben Harris.)

For example, an acoustic guitar has strings that are strung in such a way as to produce its sound. It has a resonator to amplify the sound. Sound is generated from the entire length of the strings, the inside of the sound hole, and the wood shell of the resonator. Micing just one of these components individually might not produce a balanced sound. If you simply mic the strings high on the neck, the sound will be very tinny, with little low-frequency response. If you put a mic in the sound hole, there will be an overwhelming amount of bass and overall muddiness. If you put a mic on the back panel of the guitar body, it will sound muffled, like being behind a wall. However, if you place a mic somewhere in front of the guitar so that it receives information from the strings, sound hole, and the body, you can get a beautiful, well-balanced guitar sound.

This doesn't mean that the only way to mic an acoustic guitar is to place it where we just said. No, it means that if you place it somewhere and the sound is too bright and tinny, you might want to move the mic away from the strings. If the sound is too boomy, you might want to move the mic away from the sound hole. And if it is hollow sounding, you might want to move your mic more toward the body of the guitar.

Knowing the acoustics of an instrument gives you a good idea of what might sound good, but when it doesn't sound quite right, you'll have an idea of where to go next.

Acoustic guitars can be miced very closely and have a balanced yet up-front and intimate sound. However, a lot of instruments don't deal well with being close-miced in a studio setting. As a general rule, most traditional band and orchestra instruments sound best with the microphone back a few feet. A lot of these instruments send very focused frequencies in many different directions. Some send only upper-midrange frequencies forward, high frequencies upward, and low and low-mid frequencies to the back and sides. If you put a microphone close and directly in front of these instruments, you will get an overwhelming amount of upper-mids, but if you bring it back a few feet you will capture a better balance of all the frequencies.

If you are interested, there are many wonderful drawings in books, various Websites, and tutorials showing each instrument and which frequencies are sent in which direction. However, I feel that it is better to gain a basic understanding of this concept and how to deal with the problem rather than memorizing each instrument's acoustic quirkiness. You will start to get a feel for the quirkiness of each instrument as you gain more experience recording them.

Proximity Effect

One of the problems and benefits of close micing is *proximity effect*, a natural bass boost that happens when unidirectional and bidirec-

Figure 9.5
A DJ speaking close to the microphone to give her voice more low end and presence, making it sound larger than life. (Photo courtesy of iStockphoto, Jennifer Borton, Image #2125660.)

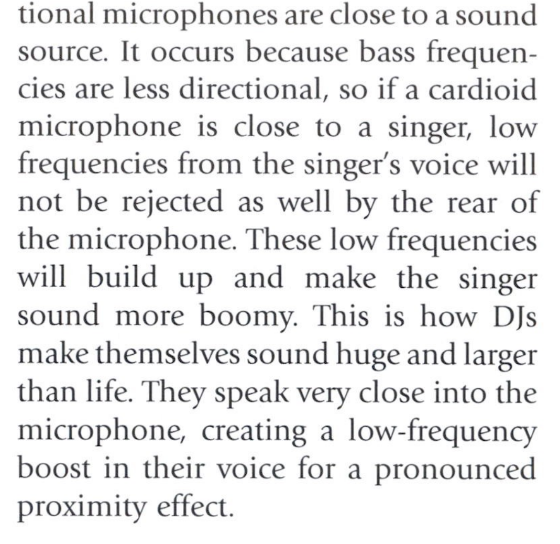

tional microphones are close to a sound source. It occurs because bass frequencies are less directional, so if a cardioid microphone is close to a singer, low frequencies from the singer's voice will not be rejected as well by the rear of the microphone. These low frequencies will build up and make the singer sound more boomy. This is how DJs make themselves sound huge and larger than life. They speak very close into the microphone, creating a low-frequency boost in their voice for a pronounced proximity effect.

Common Problems

One of the most common problems I hear in tracks produced in home studios is that the sound is boomy, with a lack of upper-mid and high-frequency detail. It took a while, but I finally figured out why so many home-produced tracks sound this way.

The first problem is that the room acoustics are usually less than satisfactory, so to take the room sound out of the equation, people move the microphone closer to the sound source. Then you only get a limited sound of the instrument, not allowing the sound to mix and develop together. Finally, proximity effect comes in to boom it all out. In addition, I have noticed that lower-quality converters and microphone preamps tend to insubstantially reproduce upper-mid and high frequencies, making instruments sound more boomy as well. This is one of the main reasons that a third of this book is dedicated to room acoustics. If you can use that information to make your room sound a little better, you can back the mic off a little and make your tracks sound better.

Figure 9.6
This image demonstrates the 3-to-1 rule, whereby if there are two microphones on a source, not in a stereo configuration, the farther microphone needs to be at least three times the distance from the source of the closer microphone. (Photo courtesy of Immersive Studios, Ben Harris.)

Another big problem that occurs while micing instruments is phase issues when you're using more than one microphone. If you put two microphones on one sound source, they are bound to pick up that signal at slightly different times. The difference in distance is what will determine the delay between the audio captured in the microphones (for example, if one microphone is 1 foot further from the sound sources, its audio content will be one millisecond later than that of the other microphone). Each track by itself will sound fine, but together there will be phase cancellations at certain frequencies or hollow sounding audio.

There are multiple solutions to this problem. First, use only one microphone on a source. This sounds simple, but using one mic can often be a challenge to get the sound you want, but in the end it will sound better than two mics with phase issues. Second, follow the 3-to-1 rule, which says that when you're using two mics on a sound source, the further mic should be at least three times the distance to the source of the first mic's distance to the source. Third, use a stereo micing technique. Many of these techniques utilize phase differences between left and right tracks to create a wide and realistic stereo image.

Stereo Micing

There are a handful of tried-and-true stereo micing techniques. Some of these are based on directionality to create an image, level differences, or phase differences, and some use a little of each. The XY technique is one of the most basic and very commonly used. This method uses two directional microphones facing 90° off axis from each other (one facing to the left and the other to the right). The mics are placed directly on top of one another so that their diaphragms are practically in the same spot. This makes them phase aligned, leaving directionality to dictate the stereo image (that is, information from the left gets picked up more by the left microphone, and vice versa). The result is a phase-accurate stereo image that is not terribly wide but is definitely stereo. If two bidirectional microphones are used instead of unidirectional ones, it is called the *Bluhmlein technique*.

The Office de Radiodiffusion Télévision Française (ORTF) technique is similar to XY but is essentially

Figure 9.7
Some of the most popular stereo microphone techniques. (Photo courtesy of Foundation Studios, Ben Harris.)

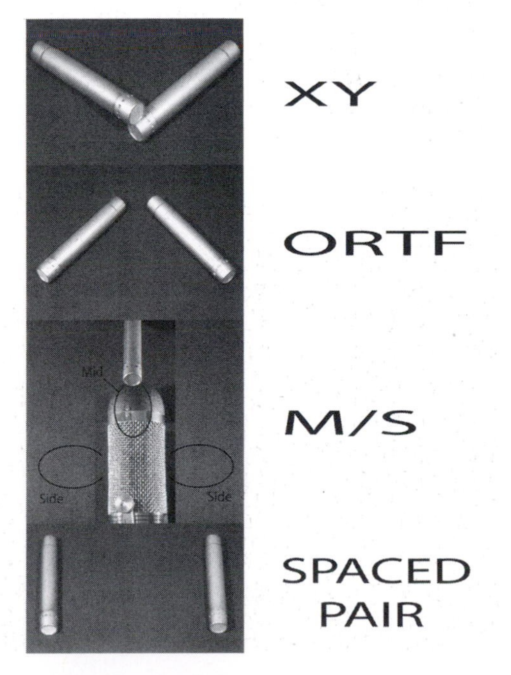

trying to recreate the position of human ears. In this method the two directional mics face opposite directions on a 90–120° angle facing out, but the diaphragms are from 6 to 9 inches apart from each other (this is roughly the distance between human ears). Since the microphone diaphragms are not aligned, there are differences in phase between the two microphones. These phase differences mimic the way human ears receive information at different times, telling us that elements are to the left or right of us. Since the mics are directional and are facing left and right, directionality also plays a role in creating a stereo image for this technique. The result is a wide and accurate stereo image with some phase instability (making mono compatibility a little more problematic).

In a variation on this technique, called *binaural recording*, microphones are actually placed in the ear holes of a dummy head so that the effects of sound moving around the head and the outer ear folds is captured, creating a very realistic 360° image when listening with headphones. However, it doesn't end up sounding terribly great through speakers.

The M/S or mid-side technique is arguably the best stereo technique and the least used. This method uses one bidirectional and one unidirectional microphone. The cardioid (unidirectional) microphone is aimed at the sound source, and the bidirectional microphone is placed directly above or below the other, facing left and right of the sound source. The tricky part of this technique, and the reason that it is not commonly used, is that you don't simply take the two signals and pan them left and right, like all the other stereo techniques. The signal from the two microphones needs to be encoded into left and right tracks.

Whoa, this sounds too crazy for me, you say. It really isn't very complicated or mythological. Many preamps have built-in M/S encoders, but what they do is simple and can be done in any DAW. First, the bidirectional microphone track is duplicated and the duplicate is put out of phase. (This is called *inverted* in many programs and can be done by any button labeled with a Ø.) These two signals are panned hard left and right. At this point it should sound pretty weird and phasey. Now you bring up the unidirectional microphone track panned in the middle. The more mid (unidirectional) that you bring up, the more narrow the image; the less mid, the wider the image. The best part is that once that mid track mixes with the "out of phase" sides, there are absolutely no phase issues and you have perfect mono compatibility. In addition, this technique produces a beautifully wide and accurate stereo image.

The spaced-pair technique can be done with unidirectional, bidirectional, or omnidirectional microphones. In this method the two microphones are placed on the same horizontal plain, with a sufficient gap between them. This gap can be anywhere from 1 foot to 50 feet. The wider the gap, the wider the image (depending on how wide the sound source is). This technique uses mainly phase and level differences between the two signals to create a stereo image. This can create a very wide and larger-than-life image, but it can also have phase issues when it's listened to in mono. Sometimes an additional microphone is placed in the middle of this arrangement, creating a Decca tree formation. This is one of the most common techniques for full orchestra micing.

Common Microphone Practices

There are millions of suggestions to be made as far as which mics to use and where to place them for different instruments. That is

Figure 9.8
Some common microphone practices used for acoustic guitar, vocals, snare drum, and electric guitar. (Courtesy of the University of Colorado, Denver, Ben Harris.)

why there are complete books on microphone technique. Closing up this section, I'll cover a few of the tried-and-true standard techniques.

Electric guitar amps are probably the easiest sound source to mic. I've tried countless expensive and exotic mics close and far, but 90% of the time a Shure SM57 as close to the speaker grill as possible, aiming perpendicular but off-centered off one of the speakers, is the best placement for a distorted electric guitar. Ribbon mics are also gaining popularity on guitar amps, and I think they sound amazing for a rich, creamy, clean guitar tone.

Drums are one of the most difficult instruments to record. This is mainly because you have to use multiple microphones, so you are bound to have phase issues and bleed. One guideline is to place a dynamic microphone 1 inch over the edge of a drum aimed toward the center of the head. This is the best way to minimize the issues mentioned earlier. Half of all recording engineers swear by using individual top and bottom mics for the snare drum. The other half just use a top mic. Either technique works; it just depends on how you hear the snare drum and what it takes to reproduce that. You usually have to flip the phase of the bottom mic, but other than that everything works out fine with either technique.

Well-placed stereo overheads are usually all you need for cymbals (the high-hat usually gets its own mic, though). Some engineers use the overheads for the sound of the whole kit, with the close mics filling in; others use overheads just for cymbals.

Another suggestion is to utilize the null of your directional microphones to help minimize bleed. Move the snare and high-hat as far apart from each other as the drummer will let you, and put the back of the snare mic toward the high-hat. Make sure that no tom mics are also aiming directly toward cymbals. Finally, when it comes to the kick drum, remove the front head so you can be free to place the mic wherever it makes the drum sound the best.

Vocals often sound really good up front and personal, so you can have the singer practically eat the microphone. If the proximity effect gets overwhelming, back the singer off 6 inches to a foot. Move the mic a little higher so it is aimed toward the singer's nose, or forward if you want a little brighter or more intense sound. Use a pop filter to avoid nasty plosives like p's and t's. Or move the mic so that it aims at the singer's cheek if you don't have or don't want to use a pop filter.

DYNAMICS PROCESSORS

A dynamics processor is any type of processor that affects the amplitude level of a signal. There are four main types of dynamics processor: gating, expansion, compression, and limiting.

A gate lowers the volume of a signal (or completely turns it off) when the signal goes below a set threshold. For example, a tom microphone picks up the rest of the drum kit when the tom is not being played. A gate can be placed on this track and set so that the gate is open (or the signal may pass through) when the tom is being played, but when the tom is not being played the audio level drops below the threshold and the gate closes (not letting any signal through and turning off the signal).

Figure 9.9
The four main types of dynamics processors: limiter (upper left), expander (upper right), gate (lower left), and compressor (lower right).

An expander does the opposite of a gate. It turns the signal up when it goes above the threshold, making the signal more dynamic. An expander can be used to make the louder parts louder, to make sections of a track pop out a little more.

A compressor is the most commonly used dynamics processor. It turns the signal volume down whenever the signal goes over a set threshold. The result is a limited dynamic range, or basically softer loud parts and louder soft parts. This processor can be used on most anything to decrease its dynamic range.

A limiter is basically a really mean compressor. A limiter has a very high ratio so that it essentially turns the signal down so much that the signal is never permitted to go over the threshold. Limiters are commonly used in mastering to process the dynamics of an entire mix. They are also used during tracking (between the microphone preamp and converter) to ensure that no loud signals clip the AD converter and cause horrible-sounding digital distortion.

Side Chain

All these dynamics processors react to a signal going above or below a threshold. In most circumstances it is the same signal triggering the processor that is being processed. A side chain is a way of sending a different signal to trigger the processor than the signal being processed. There are two types of side chains: a side-chain EQ and a separate side-chain input.

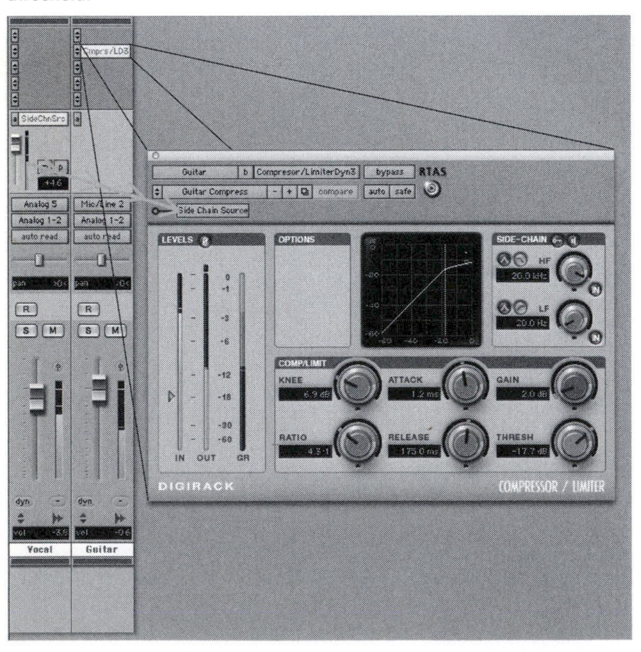

Figure 9.10
A vocal track being sent to the side chain input of a compressor on a guitar track, creating a "ducking" effect. The compressor will engage on the guitar whenever the vocal signal goes over the threshold.

A side-chain EQ takes a portion of the signal that is in the processor, adds an equalizer to cut or boost certain frequencies, and then uses that EQ'd signal to trigger the processor. What happens is, if you boost the highs on the side chain of a very punchy track, the compressor will react more on those high frequencies, resulting in a more mellow and less punchy-sounding track. A deesser is simply a compressor with a side-chain EQ boosting the problematic frequency so that the compressor reacts to just the piercing essing sound.

A side-chain input lets you trigger a dynamics processor with a completely separate signal. One technique that uses this method is called *ducking*. This is used often with a DJ speaking over a music bed. In this circumstance a compressor is put on the music bed and the DJ's voice feeds the side-chain input of that compressor. Now every time the DJ speaks, the compressor turns down, or "ducks," the music so that the DJ can more easily be heard. You can also use a side chain to tighten up the groove of a kick drum and bass guitar. The bass guitar has a slower attack and actually sits in a little pocket slightly after the big transient (or peak) of the kick drum. If you look at the waveforms of these two tracks in a good groove, you will see how the bass sound starts out small as the kick transient hits, and as the kick waveform decreases, the bass waveform rises to full amplitude. By placing a compressor on the bass and triggering it from the kick drum through the side chain, the compressor will make sure that the volume of the bass only comes up after the kick transient has finished, thus ensuring a better groove between the kick and bass.

Finally, you can use a side chain on a gate with a drum room mic to make the snare drum sound huge. Place a gate on the room mic track and trigger the gate from the snare through the side-chain input. Now when the snare hits, the room mic will turn on, making the snare sound huge. Then the room mic will turn off when other elements, such as kick and high-hat, are playing. A side chain can let you create some very interesting sounds or effects with any type of dynamics processor. Check theDAWstudio.com for detailed examples of these and other side-chain techniques.

Perceived Loudness

Compression does a magical thing: It turns down the peaks and makes the music sound louder. This is possible because of perceived loudness. Most meters in DAWs are peak meters. This means that they measure the peaks at a given moment and display that measurement on the meter.

Human hearing doesn't work like a peak meter. Humans hear in averages, not peaks. So if a track has a low signal that quickly jumps to a high peak and then back down to a low level, our brain averages out the two and we perceive the track as of medium loudness. If the same track jumps to a peak and then consistently stays at that peak level, our brain will average that consistent level and we perceive the track as very loud. In both circumstances the track never jumped

Figure 9.11
This image shows what happens to a compressed signal to make it seem louder. The peak is not any louder than the original, but the average (or RMS) level is higher, making the signal have a higher perceived loudness to the human ear.

above the same peak level, but our perceived loudness varied greatly. Compressors turn down the peaks only, then the entire signal is turned up, increasing the average levels and therefore our perceived loudness of the track.

Compression Basics

All compressors come in a variety of types; some have very few parameters to adjust, some have many. There are five basic parameters that control how a compressor acts (whether or not they are found as knobs on a processor). The parameters are threshold, attack, release, ratio, and makeup gain. On different units some of these parameters may be set and are not adjustable, some may automatically change based on the incoming signal, or they might all change with the twist of one knob. Regardless of how a compressor looks, these parameters define how the compressor acts.

Threshold is the line that, when it is crossed, tells the compressor to turn the volume down. By lowering the threshold you are telling the compressor to work against more signal content. Turning the threshold up will cause the compressor to react to only the highest peaks of amplitude.

Attack is the amount of time that the compressor takes to turn down the signal once the threshold has been crossed. This is usually mea-

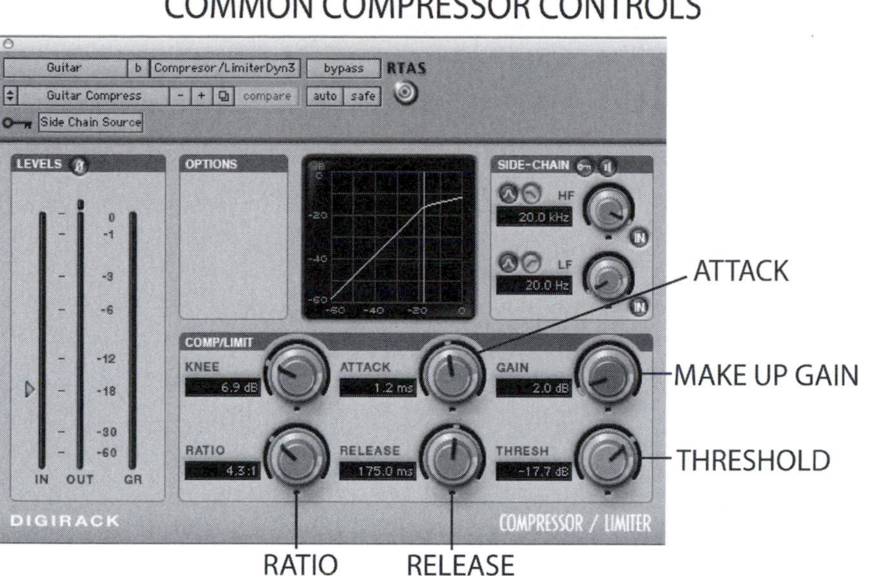

Figure 9.12
The most common controls of a compressor. Even though many compressors do not have these knobs, these parameters may be preset, combined into single controls, or called by different names.

sured in milliseconds, so if the attack is set to 50 milliseconds, once a signal goes above the threshold, 50 milliseconds later the signal will be turned down by the level set by the compressor.

Release is the amount of time that the compressor takes to turn the signal back up after the content has gone back below the threshold. This is also sometimes called *hold* because it is how long the compressor continues to hold onto the signal after it has dropped below the threshold.

The *ratio* is how far the compressor will turn the volume down in ratio to how far over the threshold the signal goes. For example, if the audio goes 3 db over the threshold and the ratio is 3:1, the compressor will turn the signal down 2 db, only letting signal 1 db over the threshold pass. If it goes 6 db over, it will let 2 db pass, and so on. Once the ratio goes over 10:1 it is called *soft limiting; hard limiting* is usually 100:1 or ∞:1.

Makeup gain is simply an amplifier that lets you turn up the entire signal once the peaks have been turned down. This is why some early compressors and limiters were called *limiting* amplifiers. If you leave a compressor's threshold at zero, you can adjust the makeup gain and use it simply as a line-level amplifier.

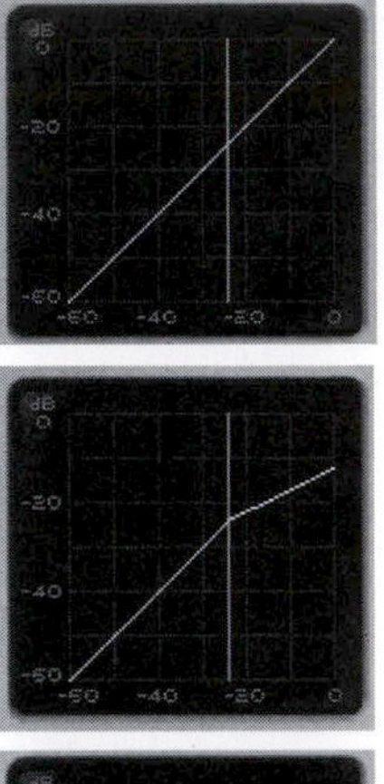

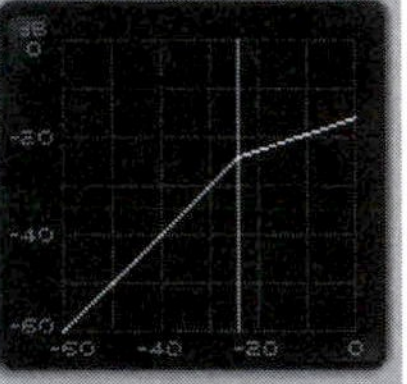

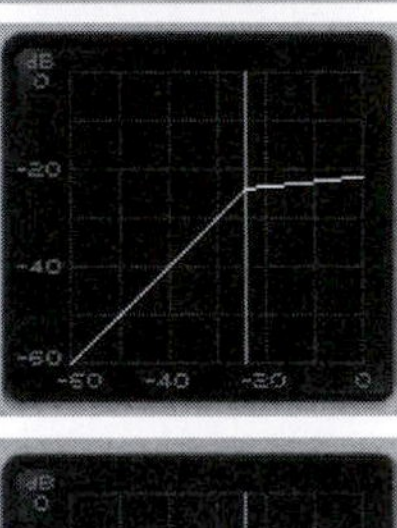

Figure 9.13
A visual representation of compression ratios. The vertical line represents the threshold, with the left axis showing output level and the bottom axis showing input level. The 1:1 ratio is a straight line because the input is equal to the output. As the ratio increases, the output level changes. The 100:1 ratio is a flat line (hard limiting) that does not let any signal over the threshold.

Compression Techniques

Compression is used for three main purposes during recording and mixing: to level the dynamics of a track, change the sonic character of a track, or create an effect. On a basic level, compression is most often used for simply calming down the dynamics of a track. There are many compressors that do a great job for this. You usually want a fairly transparent compressor that will work on the dynamics while not dramatically changing the sound of the audio content. Many stock software DAW compressors are great for these tasks. They simply do their job quietly in the background. Most of the time the goal with compression is to simply make a track louder.

A very common use of compression is to change the sonic quality or character of a track. The reason that a compressor changes the sonic character is because when it reacts to the peaks of an audio track it essentially calms down whatever is the most active (or loudest) aspect of the track. For example, if a track has a nasally upper-midrange quality, a compressor will essentially turn down those prominent frequencies when reacting to the peaks of the audio. The result is a less nasally sounding track.

Using a compressor as a sonic tool can be very effective and can often make it unnecessary to use EQ. For this reason I often try to get a

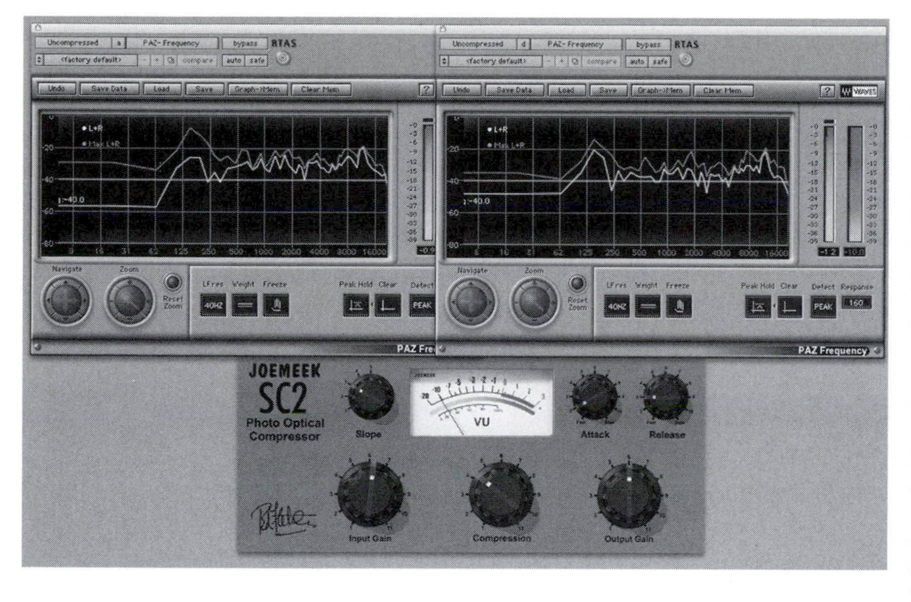

Figure 9.14
The effects of a compressor on the frequencies of a signal. The upper-left image is a frequency analyzer showing the signal before compression. The upper-right image is an analyzer after compression. The compressor is shown at the bottom. Notice the difference in the topmost line between left and right images. The compressed signal is similar to the uncompressed signal in level at 2 KHz, 8 KHz, and 10 KHz. Notice how many of the other frequencies are much lower in signal after being compressed. The result is a signal with a different frequency character without using EQ.

track where I want it with just a compressor at first (only if it is a track that I would have compressed and EQ'd). Then I might add EQ to make slight changes. Avoiding EQ in this way can help keep your mix clear, avoiding phase shift that naturally occurs when you use EQ. Changing the order of your compressor and equalizer will change how both act as well. If EQ is placed before a compressor, the compressor will react against the changes of the equalizer. So basically, if you boosted some frequencies, the compressor will turn those back down. This can be very effective in shaping a track, but it can also be simply fighting against yourself. You can order your EQs and compressors however you like, but just realize that the order does make a difference.

Compression is also commonly used as an effect. There are many ways to accomplish this task. One of the most common ways is to use it on drums to make them sound punchier and to make them breathe. This is accomplished by setting the compressor to a longer attack and shorter release. This allows the first part of the signal (the transient or big peak) to pass through unaffected. The compressor closes down on the signal as it decreases, turning down the signal's ringiness. As soon as the signal drops below the threshold, the compressor quickly releases, bringing up the tail of the decay or ringiness. The end result is a punchier-sounding track because the transient

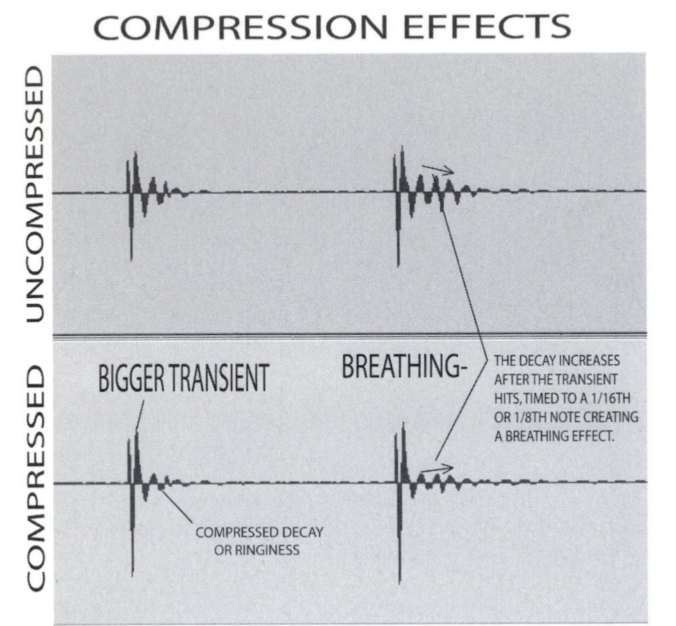

Figure 9.15
The before and after effects of two compression techniques commonly used on drums. First, on the left it shows the effect of using a slower attack time with compression. This lets the transient (or peak) pass through and turns down the ringiness. Second, on the right it shows the effect of a quick release. This brings up the ringiness rhythmically after the transient, creating a breathing effect.

becomes more apparent, with less of the loose ringy sound of the drum following. This technique is great for making drums punch through the other tracks in a mix. The breathing effect is accomplished through the adjustment of that short release time. When the quick release brings up the end of the decay and ringiness of the drum, it tends to sound like a breathing in effect. If the release time is adjusted to have the breathing occur in tempo with the song, you can get this great off-beat excitement groove happening.

Another technique that uses compression as an effect is *parallel compression*. This technique is also referred to as *upward* or *New York compression*. Using the term *upward compression* makes the most sense when you're describing it because then it can be compared to normal "downward" compression. The first step in upward compression is to either send a portion of a track off through an auxiliary send or duplicate it. The duplicated or sent track gets compressed while the original remains unprocessed. The two are added together in the mix, and the result is that the transients of the uncompressed track mask (or cover up and cancel out) the compressed transients from the processed track. Then the lower-level audio from the processed or compressed track masks the lower-level audio from the unprocessed track. The result is a signal that doesn't have any compressed transients, but the low-level audio has been turned up or brought upward. New York compression is doing this while additionally drastically EQ'ing out the midrange frequencies of the compressed track. This technique can add some body and life to somewhat flat and lifeless drums.

Mixing Considerations

There are many specific considerations to be aware of when using compression during the mixing process. These include overcompression, subtlety, and which tracks to compress.

Overcompression could be called a sickness, a disease, or a rite of passage. I feel that everyone in audio goes through an overcompression phase. The difference is that some engineers never get out of the phase. This phase is a necessary time in learning how to hear compression. Compression is difficult to hear in low amounts with an untrained ear, so most inexperienced people tend to overcompress so that they can hear the effects of their work. Start your phase now so that you can get through it quickly! The key to compression in a mix is subtlety, but you have to know how to hear those subtleties.

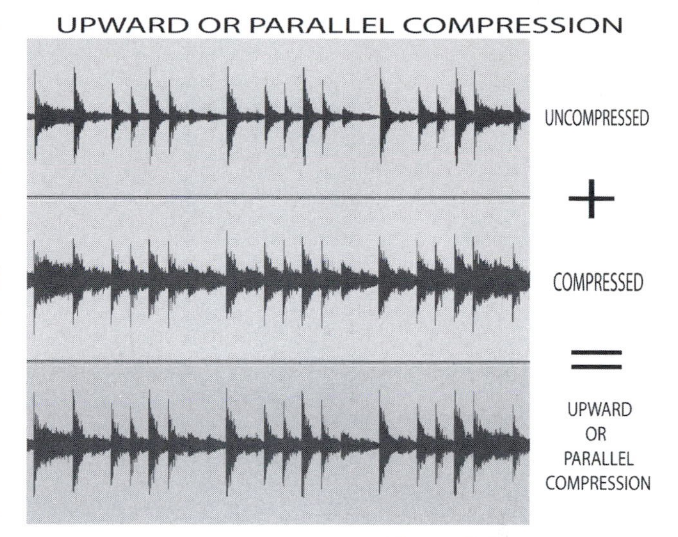

Figure 9.16
This image shows how upward or parallel compression works by adding the uncompressed and compressed signals to each other. The result is a signal with the same transient peaks as the original but with increased low-level signal. This way it pulls the lower signal up without turning the peaks down.

Once you have made it through your overcompression phase, you can focus on the subtleties of compression. I have had many experiences where I am mixing a project and I add slight amounts of compression to multiple tracks. After completing that, the client pipes up and says, "You know, I couldn't hear any difference when you turned the knobs on each track, but now the whole thing sounds better." I've even disabled all the compressors and compared it. They always can hear the differences on all the tracks, but not the subtle differences on each individual track. Most of the time compression changes are very subtle on individual tracks, eventually making a big change on the sound of the entire mix. Obviously there are exceptions, with some tracks being purposefully overcompressed as an effect or using some of the drum compression techniques mentioned earlier. The key is knowing when to be subtle and when to slam it hard.

Knowing which tracks to compress is often the most difficult aspect of mixing. Any track can be compressed, but many tracks will just sound worse if they're compressed. Any instrument that has long sustained notes with small peaks (some piano parts, harp, or any instrument with a drone) might create an undesirable sound when it's compressed. Vocals, distorted guitars, and drums can be compressed pretty hard, with excellent results. Acoustic guitar, piano, and cymbals can easily be ruined with some or too much compression. Compression is the sound of popular music, so don't be afraid to use it, but if you're not doing popular music such as classical or bluegrass, be very subtle or don't use it at all.

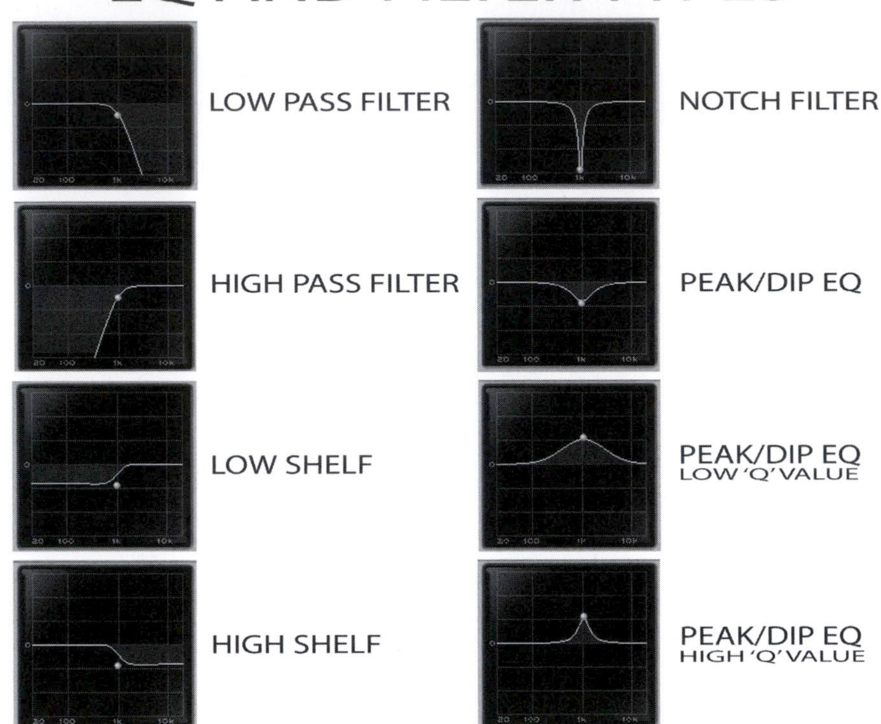

Figure 9.17
The main types of EQs and filters used in most equalizers. This image also shows the difference between low and high Q values with a peak/dip EQ.

EQUALIZATION

Equalization, or EQ, is a tool used to boost or cut various ranges of frequencies of a signal. It is probably the most common processor used in audio. Equalization is found on almost every audio console, it is incorporated into many other types of processors, and it is even in car stereos.

Filters, Peak/Dip, and Shelving

On a basic level, an equalizer can be a simple filter. A filter is anything that stops a portion of a signal from passing through. An equalizer usually consists of multiple filters to allow more control over the way a signal is processed. There are a few main types of filters used in most equalizers: a high-pass filter, a low-pass filter, a band-pass filter, and a notch filter.

A high-pass filter lets all frequencies above (or higher than) the filter frequency pass through. Therefore it cuts the low frequencies out or stops them from passing through. This is the reason that a high-pass filter is also called a *low cut*. A low-pass filter does exactly the opposite, stopping all frequencies above the cut-off frequency while letting the lower frequencies pass through. A band-pass filter is basically a high-pass filter and a low-pass filter added together to allow only a limited band of frequencies to pass through. A notch filter does the opposite, stopping a very narrow band of one or two frequencies and letting everything else pass through higher and lower than the notched frequency.

All these filters can be used individually for many useful functions, such as removing low rumble with a high-pass filter, limiting brash high frequencies with a low-pass filter, creating a telephone effect with a band-pass filter, or removing one problematic frequency with a notch filter.

There are two more types of equalizer functions that don't quite function as filters because they don't completely filter out frequencies. These types are used to boost or cut a range of frequencies. The two types are peak/dip and shelving. A peak/dip equalizer is used to boost or cut a defined range of frequencies. It is focused around, and labeled by, the center frequency. The width of the affected frequency range is defined by the Q (quality) value. This number ends up looking backward because a higher Q value causes the peak/dip to effect a more narrow frequency band, and vice versa. A shelving equalizer comes in two forms: a high shelf and a low shelf. A high-shelf equalizer boosts or cuts the focus frequency and everything higher by the same amount. A low-shelf equalizer boosts or cuts everything below the focus frequency by the same amount. These two equalizer types are used for the majority of equalization tasks.

Types of Equalizer Processors

The best thing about modern equalizers is that most have all the capabilities of the filters mentioned previously and much more. There are a few main types of equalizers, with every imaginable combination in between. The main types are graphic, fixed, and parametric.

Graphic equalizers consist of individual sliders for each frequency range. Depending on the unit, each slider will

Figure 9.18
The three main types of EQ processor: parametric, graphic, and fixed. (Photos courtesy of Immersive Studios, Ben Harris.)

affect 1/12 of an octave to a full octave each. ⅓ octave increments are probably the most common. If you pull down one slider, it works like a notch filter, or you can adjust multiple sliders to create custom equalization curves. Graphic equalizers are most often used in live sound reinforcement and are not seen much in recording studios. Fixed equalizers are found on many inexpensive mixers and processors as well as more vintage equipment. These usually have fixed frequencies and fixed filter or equalizer types. They can be useful for some situations, but many times they are very limiting.

Famous engineer and producer George Massenberg invented the parametric equalizer in the late 1970s. The parametric EQ has three parameters for each equalizer type or filter. It has adjustment of the frequency being processed, the Q value (or width of the band being processed), and the level of boost or cut. These equalizers are the most common used in recording studios. They give you the greatest flexibility of all types of equalizer processors. As mentioned earlier, there are many combinations in between the main equalization processor types, and many units are semiparametric, often having fixed Q values but adjustable frequency and gain.

The Three Purposes of EQ

The three purposes of EQ are to make a track sound better, to act as an effect, and to balance frequencies in a mix. One of the easiest ways to make a poorly recorded signal sound better is to add some equalization. With minor amounts of EQ added or taken away at the right frequencies, a muddy signal can sound clearer, a distant signal can sound closer, and a loose signal can sound tighter. The techniques for accomplishing this are very simple, and most people develop their own based on trial and error. One of the only commonly used techniques is to first do a 3–6 db boost on a peak/dip band. Second, adjust the frequency of that band until you find the worst-sounding frequencies in the signal. Third, simply pull down the gain so that you cut instead of boost those bad-sounding frequencies.

Figure 9.19
This image shows how to use low- and high-pass filters at 400 Hz and 4000 Hz to create a telephone effect.

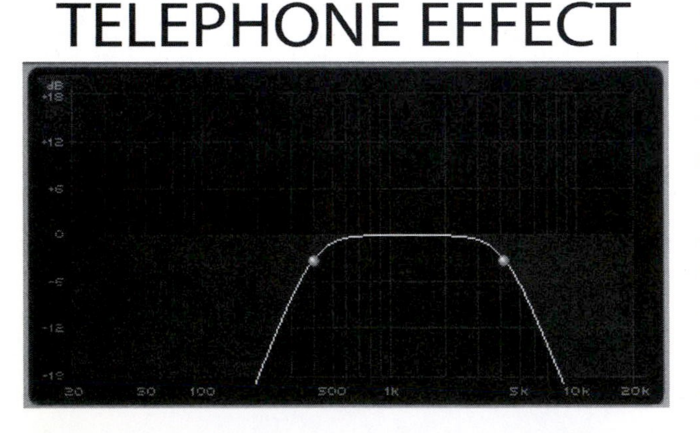

EQ is often used as an effect in a mix. Have you ever noticed how a vocal can sound small at first and then suddenly become big and full? This is often accomplished using an equalizer to make the signal sound small by filtering off both low and high frequencies. When the time comes to make the signal sound large and full, you simply disengage or bypass the equalizer. This is similar to a telephone effect, which is having high-pass and low-pass filters at roughly 400 Hz and 4 KHz, respectively. Both these techniques make the signal sound smaller, limited, and more lo-fi. You can also use only one filter to leave just the highs for a tinny-sounding bodiless effect, or leave the lows to make the signal sound like it is playing through a wall in another room. Finally there is also a filter sweep. This can be done in many ways. It can be automated with plug-in automation or it can be driven by an LFO. A filter sweep can be a band-pass, a notch, or a peak/dip EQ that moves back and forth through the frequencies while the content is being played. This effect is often used with drum loops to make them sound different from the original. A filter sweep can produce a very cool sci-fi sounding effect. You can accomplish countless effects with equalization. Just remember that many effects use a combination of both equalization processing and distortion or time-based effects.

The most common use of equalization is to use it to balance frequencies during the mixing process. The purpose for this is based on the psychoacoustic aspect of masking. When two signals produce the exact same frequency, whatever device produces that frequency at a higher amplitude will mask the other device's frequency. The best way to do this in real life is to have one person talk at a normal level, then have a second person yell something totally different over the top of the first person talking. You will notice that the normal talking virtually disappears at times when the yelling is masking it. Our ears naturally do this (which is a big part of *psychoacoustics*) so that we do not get overwhelmed by all the sounds happening simultaneously around us.

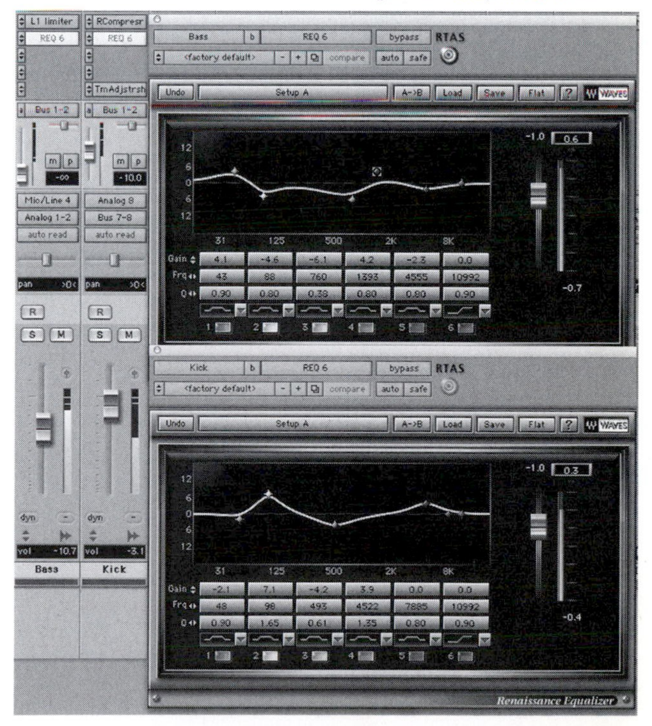

Figure 9.20
An example of balancing frequencies between bass guitar and kick drum. Notice how the frequencies boosted in one track are attenuated (pulled down) in the other.

Masking happens in any audio mix where similar frequencies from different instruments cover, or mask, each other. The best way to solve this issue is to use equalization to take away conflicting or masking frequencies. It is always better to take away more than add when using equalization. This is because of phase shifting that naturally occurs with EQ. Adding causes more phase shifting than taking away, so taking away is always better. You'll also find that subtractive EQ is more effective at fixing these masking problems. Basically, if you have a kick drum and a bass guitar that each sound great individually but when heard simultaneously sound kind of muddy, weak, and undefined, you probably have some frequencies masking each other. This is a very common problem because these two instruments have a similar frequency range. By subtracting some frequencies from one you will notice how those same frequencies pop out on the other. With these instruments in particular you might want to find one frequency to subtract on one and boost on the other, and vice versa, so that you basically have your kick drum living at one frequency and your bass living at another.

Another circumstance might be where every time that a guitar part comes in you feel that you have to turn up the vocal. You probably need to take out some frequencies in the guitar part that are masking your vocal. Once you find the frequency and dip it down a little, you will notice that the guitar part can be just as loud, but you don't have to turn up the vocal to make it heard. It is good to use the same technique we mentioned earlier to find bad-sounding frequencies to help balance a mix, but overall experience and trial and error will be the way that you will learn how to EQ effectively.

Figure 9.21
A tape machine that could be used for tape delay. The delay time is defined by the distance between the record and play heads (shown) and the tape speed. (Photo courtesy of Immersive Studios, Ben Harris.)

TIME-BASED EFFECTS

A time-based effect is any effect that varies the sound by repeating the signal on top of itself later in time. On a basic level this is simply called delay, but there is also reverb (which has thousands of delays) and chorus, flanging, and phasing (which process the delayed signal, change the delay time, or both, to create different effects).

Delay

Delay is simply a repeat of the same signal later in time. The first delays were accomplished by sending a signal to an extra tape machine. The signal would be recorded on the tape by the record head, travel a few inches, and then be played back off the play head. The signal was then sent back to the console, coming back delayed from the original because of those few inches between the record head and the play head. To have more repeats, you could send the delayed signal back into the tape machine and it would be delayed by the same amount again. This is called *feedback* (because you are feeding the outgoing signal back into the tape machine), and the more feedback you have, the more repeats of the signal. This method is called *tape delay*. This is the sound of delay on a lot of records from the 1950s and 1960s.

Then there was analog delay, which used electric circuitry to hold and release the signal by certain amounts of time. The most commonly used delay now is the digital delay. It has limitless possibilities and can be used to mimic the sound of both tape and analog delay of the past.

The key to delay is that the repeat is lower in amplitude than the original. It is the sound of someone yelling "Hello" in the mountains and hearing "Hello" repeated back. Delay is usually used in tempo with the rhythm of a song (and if the song has been recorded to a click track, it is usually very easy for a delay processor to follow the current tempo with the delay times). Often delay is used and very apparent in a mix, but most of the time it is a subtle effect, barely audible and adding some thickness to a track. Most DAWs will calculate the delay time of a quarter note for your current tempo, but it is easy to figure out based on the idea that tempo is based on number of beats per minute (60 seconds) and delay is measured in milliseconds (1/1000 of a second).

Figure 9.22
This image shows how to calculate the delay time of different beats in milliseconds based on the song's beats per minute (BPM). (Created by Ben Harris.)

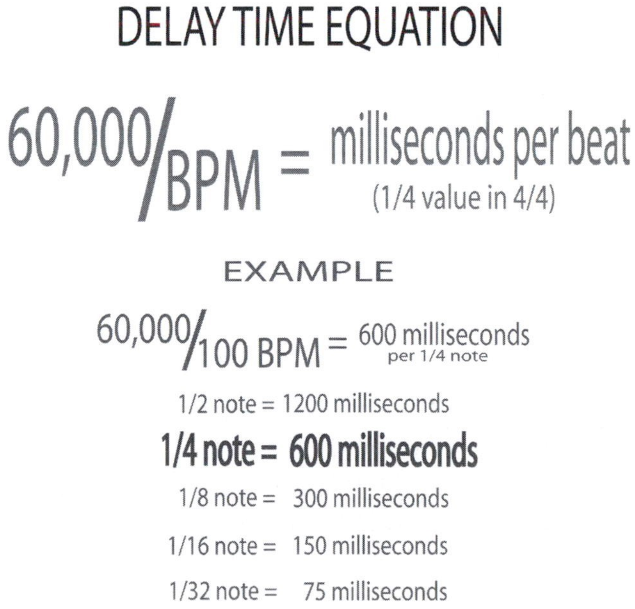

DELAY TIME EQUATION

60,000/BPM = milliseconds per beat
(1/4 value in 4/4)

EXAMPLE

60,000/100 BPM = 600 milliseconds per 1/4 note

1/2 note = 1200 milliseconds

1/4 note = 600 milliseconds

1/8 note = 300 milliseconds

1/16 note = 150 milliseconds

1/32 note = 75 milliseconds

Figure 9.23
This is what a plate reverb looks like on the outside. It's basically a box that sits away from the control room. This image shows some of the connections and how big it really is. (Courtesy of the University of Colorado, Denver, Ben Harris.)

Reverb

Reverb is a series of thousands of delays at varying levels and lengths. These delays emulate the reflections that occur in a real room. Every time a signal bounces off a surface and returns to the listener, it is heard as another delay or repeat of the signal. A handful of devices are used to generate a reverb effect to be used in recording. These effects include chamber, plate, spring, digital, and convolution.

Chamber reverb is one of the oldest ways used to generate reverb. A signal is sent to a speaker placed in a room that is dedicated as an echo chamber. The signal bounces around the room and is picked up by one or two microphones, bringing a processed signal back to the mix. The processed signal is mixed with the dry signal through an auxiliary return, and you have reverb. These rooms are not built much nowadays, but the existing chambers are still commonly used. These reverbs are heard on lots of recordings from the 1950s and 1960s. You could easily make a bathroom or large room into a makeshift chamber for a fun and unique-sounding reverb.

Plate reverb is almost as old a technology as chambers. A plate is a large piece of metal that has a speaker transducer on one end and a microphone transducer on the other. A signal is sent to the speaker transducer, it vibrates the metal plate, and those vibrations are captured by the microphone transducer and sent back to the mix. They are anywhere from a few feet to 10 or 12 feet long. They usually sit in a machine or equipment room and aren't seen from the control room. Plates sound especially excellent on vocals and are still commonly used today.

Spring reverb is similar to plate reverb except that it uses a spring instead of a large sheet of metal. These devices are most commonly found in guitar amps, but larger units are used in recording studios. You could even convert one from a broken guitar amp to integrate with your studio setup.

Digital reverb is found in rack-mount units or plug-in form. These devices create a series of delays of different time and amplitude to recreate the sound of a real space or any of the three devices we've mentioned. These units sometimes have a way of sounding unrealistic, but they create a larger-than-life effect.

Convolution reverbs are similar to digital reverbs, but they are based on an impulse response generated by actually recording and sampling the reverb of a real room. This is the most accurate reverb available and is mostly found in the form of a plug-in. These devices usually come with a large library of impulses (or samples) of real spaces, chambers, plates, springs, and even digital reverb units. This is the standard and latest and greatest technology for producing reverb. They sound completely amazing, but sometimes you might need to go back to that digital reverb to get that not-so-realistic, larger-than-life sound.

Reverb from any of these devices is usually used as an auxiliary send and return in a mix so that multiple tracks can be sent to the same unit at once. The majority of mixes have just two reverbs. There are a few guidelines that people follow for choosing these two. Some engineers advise that you find one reverb that makes your lead vocal sound good and one that makes your snare drum sound good. Others suggest finding one good reverb with a short decay time and one with a long decay time. Sometimes you will only use one reverb; other times you will use three or four. You should feel free to use one or more additional reverbs as effects on individual tracks if you

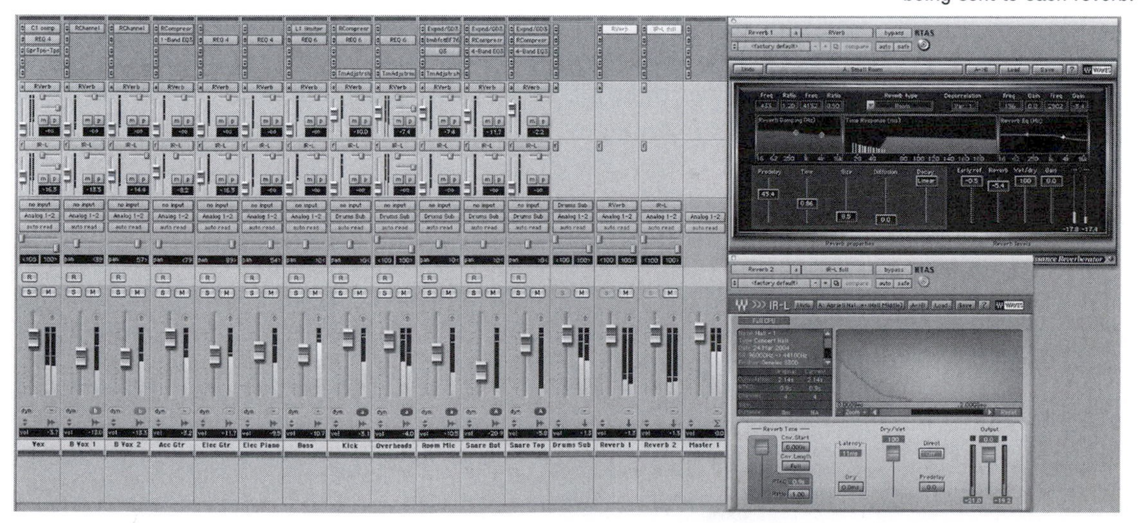

Figure 9.24
Both a convolution and a digital reverb being used on auxiliary sends and returns in a mix in a software DAW. Notice that each track has a send going to each reverb, allowing the mixer to adjust how much of each track is being sent to each reverb.

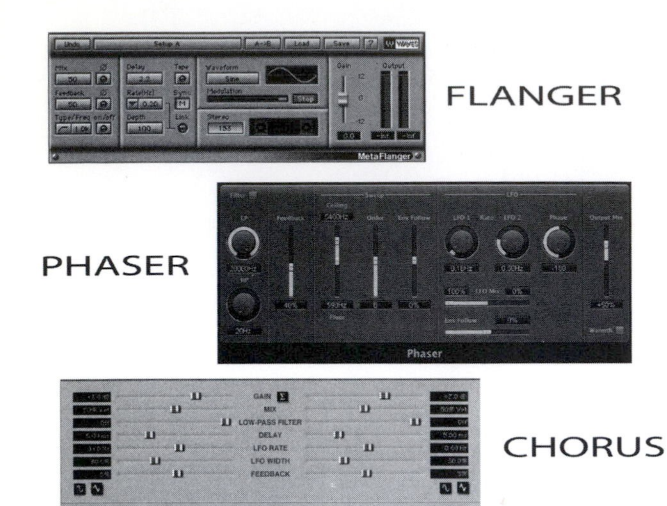

Figure 9.25
Virtual processor plug-in examples of a flanger, a phaser, and chorus.

so desire. Just don't slip into the mistake of having so many different reverbs in your song that there is no continuity tying your tracks together. Reverb is another processor that needs to be added in subtle amounts. There is nothing like a mix with huge amounts of reverb to make it sound like an unprofessional project.

Flanger

A flanger effect was originally created by playing the exact same information from two tape machines while slowing down and speeding up one machine by grabbing the *flanges*, or edges, of the tape. This technique varies both the pitch and delay of the second signal. The key to flange is that both signals are combined electronically. If each signal is sent out a different speaker, the same effect will not be heard. When these signals are added together, you get moving-phase cancellations, the thickening of frequencies, and a sweeping comb-filtering effect. A flanger can be used on any track that you want to have an unstable and moving effect. Modern flangers are varied by an LFO, so they can be synchronized to the beat of the song to create some great pulsating sounds.

Phaser

A phaser sounds somewhat similar to a flanger, but its effect is achieved very differently. A phaser splits the signal and sends one signal to be processed with a sweeping notch filter. The phase shift created by the notch filter in addition to it moving is what creates the great phasing effect. Phasers are most often used as guitar effects, usually with guitar pedals, but it can also be done in the studio.

Chorus

Chorus is an effect used to make one person singing sound like a large chorus of people. It is commonly used on vocals, guitars, and electric piano. Chorus is created by delaying a signal and then varying the pitch of the delay. The parameters of chorus generally let you adjust the rate of the LFO that is varying the pitch and the depth (or amplitude of the LFO), or how far the pitch is varied. Increasing both of these can create some crazy-sounding effects. A stereo chorus simply sends the unprocessed signal through one speaker and the processed signal through the other. Chorus is often used subtly to thicken up a track or in extreme to create a somewhat psychedelic effect.

CHAPTER 10

Bringing It All Together

The final steps of a project are the culmination of all the skills we've discussed so far. First, the project needs to be mixed, incorporating many of the tools discussed in the previous chapter. Second, the mixes need to be mastered, processing the mixes and putting them in order for an album. Third, the project needs to be presented to the world, either on CD/DVD, the Internet, or any other way.

MIXING

Mixing is the near final process of adding all of the track elements together. It is sometimes mistakenly called post-production (since tracking is referred to as production), but in audio the term post-

MIXING

Figure 10.1
People mixing on large-format analog consoles. (Photo courtesy of iStockphoto, Anna Pustovaya, Image #4784600; Immersive Studios, Ben Harris.)

production more commonly refers to providing audio for video and film.

The Purpose of Mixing

The purpose of mixing is to add all the tracks together so that each is individually audible, so that the processing of the mix supports the feel and excitement of the song, and so that the mix becomes an artistic performance. All elements of a good mix should be audible so that if you focus on any one part you can hear it when it is playing in any portion of the song.

Many home studio users don't see this clarity and separation as important because they don't hear it in professional mixes. I had always heard music over the radio, through middle-quality speakers, and in acoustically poor listening environments. My idea of what music sounded like was based on these experiences. When I began to mix my projects I mixed them to this standard (gritty and powerful, with things pushed up front and in your face) until I began to listen to professional music in a professional recording studio. Then I found out what professional mixes really sound like (pure, clear, and clean while being punchy and powerful, with depth and clarity). This was the first step to training my ear to a professional standard. After that I continued to listen to professional music on professional speakers while I mixed and while I listened for leisure.

If you have studio monitors, listen to the music that you want to make your music sound like through those same speakers. Listen to your professional reference tracks while you are mixing your project. Focus on each individual instrument and listen to its sound quality, its depth, and its level in the mix. Everybody has to do this somehow. Many professionals of years gone by went through this learning process between getting coffee, wrapping up cables, and going out for pizza as interns at professional studios. Find your way and train your ear to a professional standard.

A mellow song about love and loss should not have a sharp-sounding buzzing guitar part that sounds uneasy and biting. It should have a mellow, smooth guitar part that supports what the song is talking about. A lot of this is done in the producing and arranging process, but things can easily be fixed or messed up during mixing. There are many simple ways to control how sharp you make elements sound with equalization, how crazy the delay is, or how hard a compressor smashes a signal that can customize the sound of the mix to match the feeling or emotion of a song. As you get a better feel for mixing, you will sense the direction that a mix is going and whether it supports the song the way it should.

Another key aspect of mixing and producing is making the recording sound as exciting as a live performance or larger than life. The interesting thing about a live performance is that it is impossible to capture the same dynamics and energy of a live performance in a recording. Specific production techniques are used to add excitement to recorded tracks to make them sound more dynamic or energetic. This is the reason that recordings often have additional guitar parts and little effects in spots. Four instruments playing a song straight through can often get boring on a recording, so subtle elements are added to make the chorus sound bigger (without actually being louder) or to make the singer seem as though he or she is dancing around the stage.

Some of these techniques include automation—moving parameters in a mix in real time as the mix plays. This can include volume changes, panning changes, and processor parameter changes. Automation is the way a part pans back and forth from one side to the other during a song. This was originally done by having multiple engineers perform these automation functions manually on a console as the mix was mixing down. Then consoles began to incorporate more and more automation features, saving your moves and playing them back in sync with the mix. Now DAWs can automate nearly any function or feature in the system. The possibilities are limitless.

Figure 10.2
A piano player pointing out that, similar to performing on the piano, mixing requires skill, technique, and emotion. (Photo courtesy of iStockphoto, Robert Rushton, Image #2222173.)

Many people wonder why there are famous mix engineers in the music industry. Why do some artists or groups seek out a specific mix engineer? How can mixing make such a big difference in the final product? Wouldn't any professional engineer do a good job? My answer to these questions has most effectively been a question. Have you ever heard a pianist perform where they could play all the notes perfectly but something was missing in the feeling and the emotion of the performance? What is the difference when you get a pianist who can move her fingers right *and* convey the emotion?

It makes all the difference in the world. This is nearly identical to the mixing process.

Just like playing the piano, there are many technical aspects to mixing—aspects that will make a mix fall on its face if the engineer is not doing things technically right. These are elements such as balancing frequencies, obtaining clarity between instruments, and leveling out the dynamics. Then there is the artistic aspect, projecting the emotion and feeling of the song. Defining the way a mixer does this is as easy as explaining how a pianist does it during a performance. It's not easy at all. Sometimes it is choosing not to make frequencies completely balanced, to create tension; sometimes it is making instruments mush together a little bit, to create doubt; and sometimes it is letting something be a little more dynamic, to create some excitement. Just like the pianist, once you have mastered the technical skills, your artistic side can be free to flow.

Setting Up a Mix

The first phase of mixing is setting up a mix. Before digging too heavily into EQ'ing, compressing, and adding effects to tracks, you

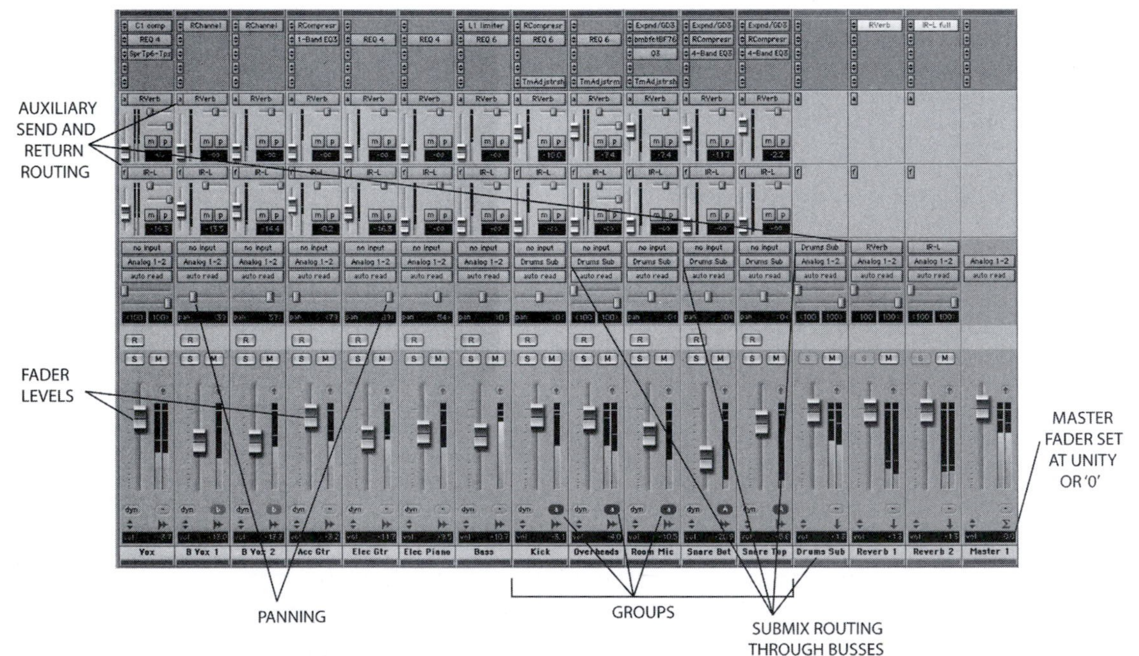

Figure 10.3
This image shows a mix, pointing out the elements that are set up at the beginning of the mixing process, such as panning, levels, groups, busses, and auxiliary send and returns.

need to do some basic routing, leveling, and panning. Some of the routing that can help set up a mix is to create a few auxiliary returns for reverb and delay with corresponding auxiliary sends on every track. This makes it easier to try reverb on each track without having to pause and do some routing in the middle of a mix.

Many mixers like to use submixes and/or groups so that all the drums or vocals can be processed or changed simultaneously. To create a submix, all the tracks you choose need to be sent to a bus in the DAW. That bus needs to be assigned to the input of an auxiliary or master track so that a processor can be inserted or level changes can apply to just the submix. A group is similar except that it involves simply pairing tracks together so that level changes on one will force all to move simultaneously. There is no special routing of the output, so you cannot place a processor on all the elements simultaneously (unless you have a submix) of a group.

You can also create memory locations or markers to let you quickly jump to different parts of a song, such as the song start, chorus, or last verse. There may also be tracks to name and reorder so that everything is in a logical place for you during the mix.

While setting up a mix, you should set some preliminary levels. Levels will be adjusted throughout the mixing process, but setting levels now will help you get a jump on your mix. First, be aware of distortion through every point of your gain stages. These tracks are all adding up together to go out the master bus. If the levels are pushed up fairly high, they will add together to be bigger than the master bus can handle and you will end up with distortion. Almost all your tracks should be below unity gain (the point at which the signal is neither amplified or attenuated or 0 dB), with the kick drum and some of the other far left and right elements being the highest at around –3 to –6 dB. This will give a good amount of headroom (the additional gain available before clipping) on your entire mix, because once you start EQ'ing and compressing your tracks, that headroom will start to disappear. A good general rule that I like to follow to maintain good headroom and avoid distortion is to have a master fader, but always leave it at unity gain (0 dB). By not adjusting the level, I can always see that, if I am clipping I need to turn my mix down, and if I have plenty of level left on the meter (or headroom) I can rest easy. Every DAW is different, but most plug-ins and tracks have meters to let you

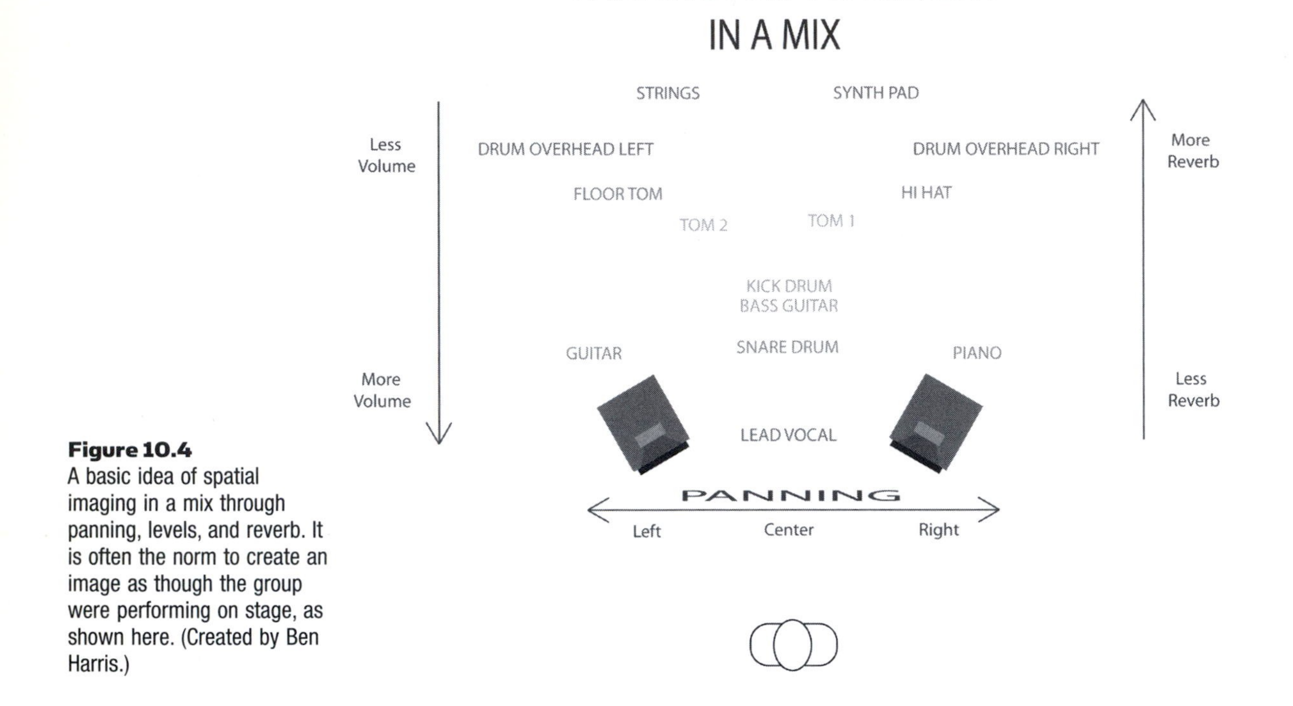

Figure 10.4
A basic idea of spatial imaging in a mix through panning, levels, and reverb. It is often the norm to create an image as though the group were performing on stage, as shown here. (Created by Ben Harris.)

know whether their input is clipping. If there is any clipping, you should go to the last step in the signal before the clipping and turn the signal down. If you stay aware of your levels (and avoid clipping) at every point of your mix, between plug-ins, through auxiliaries, and on the master buss, you will have a cleaner and clearer-sounding mix.

Panning

Panning involves sending more or less of a signal to left or right speakers in a stereo system or any of the surround speakers of a surround-sound system. One misconception is that tracks need to be stereo to be heard in stereo, but the majority of tracks are mono (one track), such as the output of a guitar, a voice, or a single drum. Most of these elements are recorded with one microphone or one input to a mono track. From this point, the output of the mono track is stereo, sending the signal equally to left and right speakers (being panned center) until the panning is moved, sending the majority of the signal to either the left or right speaker. During mixing, a mono signal can be placed anywhere in the stereo image that you desire, hard left or right, partially left or right, or dead center.

Stereo tracks are recorded from a stereo source, such as the two outputs of a keyboard or two microphones capturing the overhead sound of a drum set. If you have a stereo track where both left and right meters are identical, you have a mono source that is doubled to left and right channels. This doesn't make it sound fatter, bigger, or stereo. It just makes it sound like a mono signal being panned center. Most DAWs will let you adjust the panning of each side of a stereo track. Stereo elements do not always have to be panned hard left and right. They can be pushed inward or more toward one side or the other.

Some general panning conventions are used in popular and traditional music when working in stereo. These apply generally to how you usually pan a drum set or ensemble, what usually goes in the middle, and what usually goes on the sides (far left and right). A drum set is usually panned as though it sits in real life. The biggest issue is from whose perspective, the drummer's or the audience's? The answer is both. There is a 50/50 split among engineers to pan the drums to the drummer or audience perspective.

Let's explain it from the audience perspective to make it easier, for readers who aren't drummers. The overheads are usually panned hard left and right (unless you want to squeeze the whole image in a little bit, which is done quite often, panning them $\frac{3}{4}$ of the way left or right), high-hats are panned $\frac{2}{3}$ to the right, next comes the high tom (if there is one) $\frac{1}{3}$ to the right, next in the center is both the kick drum and the snare drum (although some engineers like to put the snare slightly off center to the left, to be more realistic, but most like it to be anchored in the center), $\frac{1}{3}$ to the left is the next rack tom, and $\frac{2}{3}$ to the left is the floor tom. If there are more or fewer toms or mics, spacing should be dealt with accordingly. An ensemble or orchestra is similarly panned to the perspective of the audience.

A few elements are usually always panned in the middle. First is the lead vocal; it is the center of attention and needs to be the focal point in a mix. Second are kick drum and snare drum; they are another main focus of any popular music song and need to be anchored in the middle to lock in the groove. Third is the bass guitar or synth bass, actually the key to any good song, setting the groove just as much as the drums. The kick and bass (as well as most other low-frequency content) usually work best if placed in the middle. Putting one of these low-frequency elements to a side really throws off the balance of a mix, and it sounds really weird on headphones, too. Lead parts, such as instrument solos, can often be placed in the center but are usually found hanging out on the edges.

Basically everything else in a mix can go out on the left and right edges. Guitar parts, keyboard parts, string parts can all be placed wherever you like. The traditional convention is to place every track as though it were a real performance on the stage. So, if the guitarist usually stands on the right, pan him to the right; and if the piano is usually placed on the left side of the stage, pan it to the left. When people first start to pan they have a tendency to keep everything toward the middle, but you should go crazy—pan things hard left and hard right. Make your stereo image as wide as possible, because it only gets smaller from here on out. Everything from playing a song over the radio to compressing it into an MP3 all make the stereo image more narrow, so make it as big as you can now.

Common Problems and Solutions

Once everything is set up, the mixing process begins. Now it is time to utilize all the tools discussed earlier, including equalization, compression, and time-based effects to make your mix sound alive and balanced. However, you are bound to run into some problems, so here are some solutions.

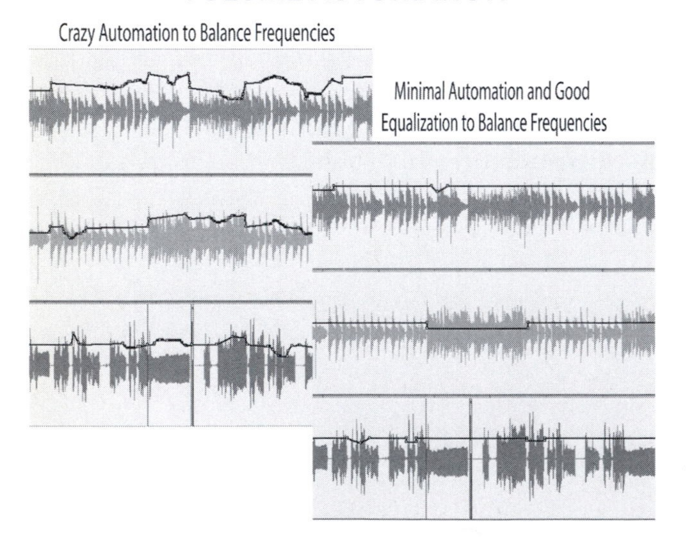

Figure 10.5
The image on the left shows how a mixer with poor EQ technique has to make the tracks fit together by constantly adjusting the volume. The image on the right shows how the same mix with good EQ balancing can easily be mixed with a minimal amount of volume automation.

One of the most difficult tasks in mixing is balancing the frequencies with equalization. One crutch that is often used is automation to make up for inefficient frequency balancing. What happens is that as frequencies mask each other, turning up the volume of the track being masked will temporarily solve the problem, but only until the content of the other track comes down a bit, revealing that the track that you just turned up is now too loud. Many people play this automation game, turning tracks up and then down as they conflict with frequencies in other tracks. The result is hours of work to not fully solve a simple problem.

The simple and correct solution is to balance the frequencies (with equalization) between the two or more tracks in question. Yes, this is difficult, and it takes time to get good at it. So use this as an indicator to let you know when you have effectively balanced the frequencies in a mix. If you feel like you need to automate every other word of a vocal, you probably need to work on EQ instead of automation. I had this exact problem when I first started mixing. Now I can pull up my old mixes, erase all the automation, and in an hour of EQ'ing listen to a well-balanced mix without any need for volume automation. I still do some automation because you can't always get a compressor to level out the most extreme and dynamic vocals, and sometimes you want a part a little louder during one section, but mostly you should use automation to create cool effects with changing levels, panning, and plug-in parameters.

Another common problem that was touched on earlier is adding too much reverb or delay to tracks. This is often a sure sign of unprofessional mixes. Many people tend to add too much reverb to a vocal to make up for pitch problems or because *they* are the vocalist and they think it makes them sound better. (Mixing your own vocals is always tough.) Many professional mixes sound like they have little or no reverb, but when you mix with no reverb it sounds just plain boring. The solution is to find a reverb that sounds good at loud levels on your track and then bring it down slowly until it is barely audible. Sometimes bring it down so that it is just past the point of being audible. Then listen to your mix with the inaudible reverb and then with the send bypassed. You will be surprised to find that it often adds thickness and richness to the mix without adding any noticeable effect. Try doing the same thing with delay and you should be equally happy with your results.

Many people solo a track and make it sound wonderful using EQ. When they listen to it with the mix, it is too thick or doesn't work at all. The solution is to do the majority of your processing while the entire mix is playing. Feel free to solo a track or a few tracks at a time to focus in on what you are doing, but don't lose sight of the big picture and go more than 10 or 15 minutes without listening to the entire mix again.

Have you ever had somebody ask you to listen to their latest mix for review, and when you start listening to it, they pipe in, "Oh, you need to turn it up because it only sounds good when you turn it

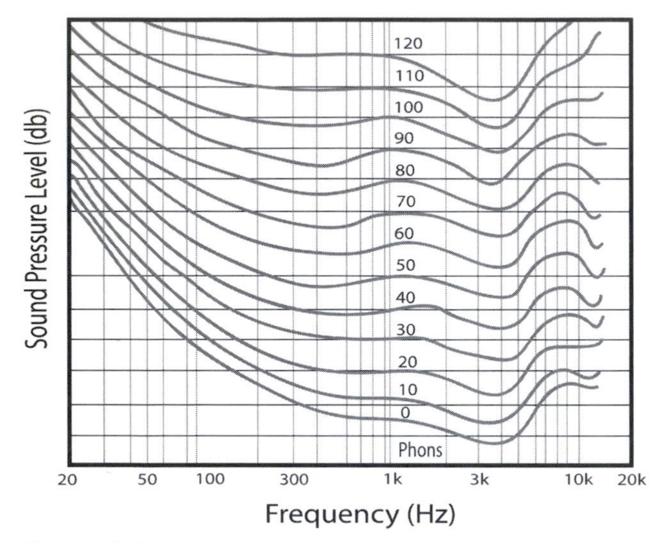

Figure 10.6
The Fletcher-Munson Equal Loudness Curve, which basically points out that it requires more energy at lower and higher frequencies to equal the loudness we hear at midrange frequencies. These levels are defined by the middle numbers of phons. (Created by Ben Harris.)

up"? Have you ever noticed that professional mixes sound good at both soft and loud levels? This all has to do with the balance of the midrange frequencies in a mix. This is because of the equal loudness (or Fletcher-Munson) curve. This curve says that our ears hear midrange frequencies more easily than low and high frequencies. Therefore it takes more energy to produce low and high frequencies at the same apparent loudness of midrange frequencies. So basically, as audio gets softer, we hear a more pronounced midrange and less of the lows and highs. As it gets louder we hear more of the lows and highs and less midrange. Try it. You will notice that as you turn down a song, the lead vocal will become more apparent and out front in the mix because it is focused in midrange frequencies. Listen to the lead vocal as you turn up the song. Other elements, such as the drums and bass (more focused in the highs and lows), will seem to start taking over and push the vocal into the background. Remember that friend whose mix only sounded good loud? It is because he mixed it loud, so he never had to critically balance the midrange. Listening loud makes it difficult to hear the midrange and easier to sound good, but when the song is listened to at a low volume, the problems become apparent. The solution is to mix at lower levels (so that if someone is talking you can understand them over the mix) the majority of the time, but turn it up now and then to check the low and high end. If you get your mix sounding good using this method, it will more likely be a good and balanced mix, and it will sound good both soft and loud.

You might have noticed that a lot of the solutions to mixing problems are to use effects and processors very subtly. I think that subtlety is the key to professional mixing. A little bit of EQ here and a little bit of compression there eventually adds up to a great mix. The problem is that until you train your ear to hear those subtleties, it is very easy to overprocess in your mixing. Consult the companion Website at theDAWstudio.com for further ways to train your ear and receive constructive feedback on your mixing.

MASTERING

Mastering was originally simply the process of transferring the master tapes to the master record to be pressed and duplicated. The technician had to check for phase issues and make sure that the levels were not too loud or the lathe would cut a hole through the record. Some technicians started to add a little more compression and EQ to make the record sound better, and now mastering is an integral step in the music-making process.

The Purpose of Mastering

The purpose of modern mastering is still to prepare the album for duplication, but it's more to make all the songs on an album sound even dynamically, similar sonically, better overall, and true to the original on any playback system.

Just like in the old days, it is the mastering engineer who creates a master disc ready for duplication. This is a little higher quality than simply burning a CD on your computer. A red-book CD is created with a minimal number of errors. If there are more errors than the set minimum, it is thrown out and a new one is created. Many mastering houses are prepared to create DVDs and any quality compressed format desirable, such as MP3, WMA, or AAC.

Figure 10.7
Air Show Mastering, a professional mastering facility in Boulder, Colorado. (Photo courtesy of Air Show Mastering.)

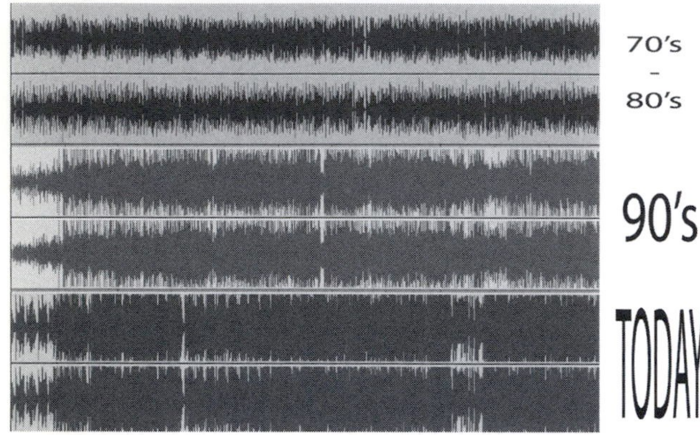

Figure 10.8
Waveforms from three songs, demonstrating the progression of the loudness wars over the past 40 years.

One of the many misconceptions about mastering is that the job is to make the music as loud as possible. It is true that many engineers are forced by their clients (especially record labels) to compress music to within an inch of its life to compete in the loudness wars, but most mastering engineers do not want to do that at all. One of the purposes of mastering is to make each song have similar levels so that one song isn't really loud and the next song is super-soft. The idea is to make the songs flow nicely so that listeners don't have to adjust the volume of their stereo systems. The engineer can also use compression to make the mix sound more exciting or punchier. Loudness wars between record labels have been going out of control since the late 1990s. Most popular music releases are compressed so badly that they have distortion throughout. This distortion fatigues your ears so that by the time you finish listening to a recently released album you want to turn off the music and rest your ears for a bit.

Another purpose of mastering is to make the songs on an album similar sonically. Songs of different sonic character can be just as jarring as songs of different loudness. Many albums are now worked on all over the world. The final mixes will have been done in three different studios on two continents. It is the mastering engineer's job to make the songs sound like they were all done in one place and were meant to fit together. The engineer may use compression, equalization, or even tape machines to accomplish this task.

If you talk to any mastering engineer, his or her first response to a question about what the purpose of mastering is will be "to make it sound better." One of the greatest benefits of having a different mastering engineer from the engineer who mixed the project is having a different set of ears to bring out new aspects of the project. When used effectively, mastering tools can help bring out excitement in a mix, make a vocal more intense, give a snare a little more crack, or make a flat song more dynamic. The goal in mastering is to accomplish all the other tasks while first and foremost making it sound better.

The last purpose of mastering we'll mention here is to make the songs sound good when they're played back on any medium. Mastering engineers have much more detailed and critical speakers than regular recording studios. By critically listening and processing the signal through these speakers, they can make an album sound good on any sound system. It is also very common for a mastering engineer to have small consumer-level stereo speakers to check playback on the lowest-quality systems. All the mastering tools help accomplish this purpose of the mastering process.

Tools and Techniques

The main mastering tools are very similar to recording tools. They consist of equalizers, compressors, reverb, and special computer software. The only difference is that the processors used for mastering are much higher precision and a lot more expensive.

Equalizers used in mastering are very clear and precise. One of the latest buzzwords for plug-in equalizers used for mastering is the "linear phase" EQ. This type of EQ corrects the phase shift that naturally occurs in any equalizer. These processors sound amazingly precise and detailed. A lot of EQ adjustments made in mastering are very slight, so many hardware units have ½ db and 1 db notched adjustments. This provides excellent recallability and very precise adjustment. In mastering there will commonly be 1 db of gain added at 5 KHz to bring out the crack of the snare, 1 db taken away at 200 Hz to clear a little mud, or 1 db boosted at 10 KHz to bring out the air in a vocal.

Compressors and limiters used for mastering also vary greatly from those used for recording. First, these units are very precise and have a fast response. They can't be a squishy optical compressor. Second, the limiters are usually brick-wall look-ahead limiters. This means that they do not let any content through over the threshold and they can see what is coming next, so they can be ready to smash anything. Third, there are multiband compressors, which we have not yet discussed. These units can be used in recording but are more often found in the mastering process. A multiband

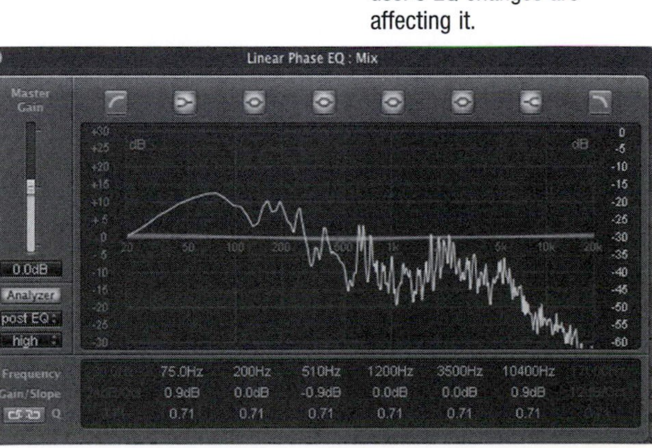

Figure 10.9
A linear phase EQ plug-in with a built analyzer, allowing the user to see what the original frequency of the signal is like and how the user's EQ changes are affecting it.

Figure 10.10
A multiband compressor, which compresses four individual frequency sections of the signal.

compressor lets you compress various sections of frequencies individually. So you can compress the lows differently than you compress the highs. This is very useful in tightening up the bass while not choking the high end.

Many mastering engineers also use tape machines for compression. Depending on the project, they will send a signal to a tape machine and push it hard onto the tape. The tape machine naturally distorts and compresses the signal in a very pleasing-sounding way. This technique can add a lot of character and thickness to a mix. Many mastering chains have multiple compressors and limiters to have a series of compression stages on a signal. This is one of the ways to completely smash a signal beyond having any life left.

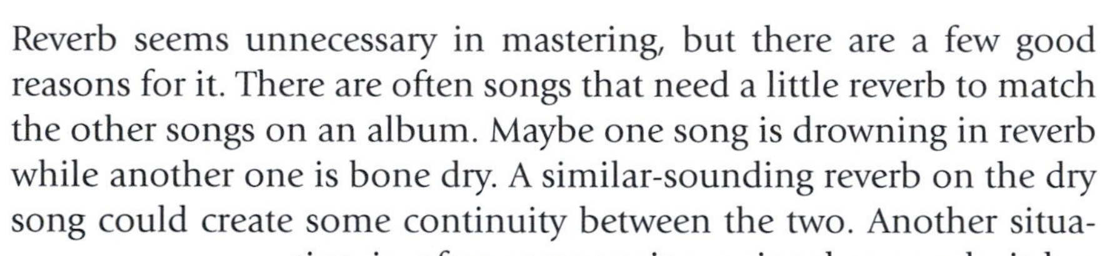

Reverb seems unnecessary in mastering, but there are a few good reasons for it. There are often songs that need a little reverb to match the other songs on an album. Maybe one song is drowning in reverb while another one is bone dry. A similar-sounding reverb on the dry song could create some continuity between the two. Another situation is, after compressing a signal so much, it has a tendency to squish down the reverb a little. In these circumstances some extra reverb can breathe new life back into the mix. Mastering engineers usually have high-quality reverb units similar to those found in recording studios. This lets them dial in the same sounds used in the mixing process.

Figure 10.11
The remote control for a high-end reverb used for both mixing and mastering. (Photo courtesy of Immersive Studios, Ben Harris.)

Most aspects of mastering can be done in DAW programs used for recording and mixing, but a handful of tasks need a special mastering application. These software applications are two-track audio editors that allow an engineer to edit and process two-track mixes. They also let the engineer add track markers and edit text and internal CD data, and they burn red-book CDs directly from

Figure 10.12
A mastering software program called Waveburner, made by Apple. This program lets users process audio, create track markers, and burn red-book CDs directly from the application.

the program. Many plug-ins and some programs advertise that with this product you can master your CD, but I would suggest going to a professional for mastering. The most important aspect and tool of the mastering process is the critical ear of the mastering engineer. Some affordable mastering services are suggested on the Website, at theDAWstudio.com.

INTERFACING WITH THE WORLD

Burning a CD

Once your project is complete, you will want to share it with others. As mentioned earlier, I suggest using a mastering service to create a quality finished product. However, many times you just need something for your friends to listen to. In this circumstance you might add a little bit of limiting to the whole mix and then burn it to a CD. There are many CD-burning programs available. Most of these programs burn orange-book CDs. The biggest difference between a red-book and an orange-book CD is that an orange book has more errors. These errors will cause your CD to have a more difficult time playing in many players. There are some inexpensive programs that will burn red-book CDs.

Figure 10.13
A burning application called Toast by Roxio. This program allows the user to create data, audio, or hybrid discs on nearly any optical media.

Sharing on the Internet

Many of you will never burn a CD of your projects. Many and probably all of you will use the wonderful opportunities provided by the Internet to interface your music with the world. In this circumstance it is better to convert your master directly to a compressed format such as MP3, WMA, or AAC. From this point there are countless possibilities of ways to share your music. The standard business model in the music industry focused around the big record label is constantly changing. Be creative and find your own way to interface your music with that big world at your fingertips. Since this is an ongoing and constantly changing topic, please consult the Website at theDAWstudio.com for up-to-date information on interfacing your music with the world.

CONCLUSION

Here at the conclusion of this book, my hope is that you have gained an excellent knowledge base that will help you understand how sound acts in your room, better be able to record it, know how to mix and process it so that it sounds like it does in your head, and be able to effectively purchase and use the equipment needed to assist you in this process. Another hope is that you will have acquired a sufficient base of knowledge and glimpses into various aspects of audio that you will be able to add more advanced books to your library that further expound on individual topics touched on in this book. Finally, feel free to utilize the excellent additional resources provided on the companion Website at theDAWstudio.com keeping current with new technologies, unbiased reviews, and current issues.

Figure 10.14
A MySpace page for a successful independent band. Notice that music can be demoed and sold directly from the site. (Screenshot courtesy of Goodbye Champion.)

Index

Page numbers followed by f indicate figures.